"It is painfully rare for a [...] alike. Aalbers and Dolfsma detail how communication and network gaps and blockages within organizations derail innovation. With rare but welcome equal billing for formal and informal structure alike, and a keen eye for how they can fruitfully combine, they show how alert managers can leverage the tools of network analysis to create an 'innovation engagement scan' that will jumpstart innovation by putting people together who ordinarily do not communicate. A seminal contribution and an indispensable aid for firms in search of innovation."

Mark Granovetter, Professor, Stanford University, USA

"The reality of organisational life is that we operate through networks of relationships. To make sense of how these networks operate, Aalbers and Dolfsma have put together this impressive book – it explains the theory of networks in a clear and accessible way, and it also shows how these academic concepts can be applied in a practical way."

Julian Birkinshaw, Professor, London Business School, UK

"In a time in which we expect big shifts for many established organizations, the navigation of social networks focused on innovation has become a core competence. Although the importance of networks is recognized, many boards and managers still have limited insights and tools to analyze, influence or even create these networks. This book offers a great opportunity for all of us to come up to speed in this highly interesting domain."

Pieter Hofman, Partner at Deloitte Consulting, the Netherlands

"Effective networks are critical to successful innovation. This valuable contribution demonstrates the significance of networks and reveals how they can best be managed."

Mark Dodgson, Professor, University of Queensland, Australia

"In an era of increasingly connected organizational networks and maturing big data analysis, executives and managers now have additional tools to design and manage the organization for innovation. *Innovation Networks* offers a fresh approach to analyzing and implementing organizational networks through a unique blend of practical methods and examples."

Li Feng Wu, Head of Analytics PayPal, China

"Networks in organizations are crucial for successful innovation. This book will help you understand how formal and informal networks, built up from relationships between people, can boost a firm's innovative power. I am convinced effective networks can improve all functions in an organization. It is all about connecting from person to person and sharing ideas and information amongst each other."

Paul Poels, Director of Digital Analytics, Philips, the Netherlands

INNOVATION NETWORKS

Organizations are complex social systems that are not easy to understand, yet they must be managed if a company is to succeed. This book explains networks, and how managers and organizations can navigate them to produce successful strategic innovation outcomes. Although managers are increasingly aware of the importance of social relations for the inner-workings of the organization, they often lack the insights and tools to analyze, influence or even create these networks.

This book draws on insights from social network theory; insights sharpened by research in a number of different empirical settings including production, engineering, financial services, consulting, food processing, and R&D/hi-tech organizations, and alternates between offering critical real business examples and more rigorous analysis.

This concise book is vital reading for students of business and management as well as managers and executives.

Rick Aalbers is Assistant Professor of Strategy and Innovation at Radboud University, Nijmegen, the Netherlands.

Wilfred Dolfsma is Professor of Strategy and Innovation at the University of Groningen, the Netherlands.

INNOVATION NETWORKS
Managing the networked organization

Rick Aalbers and Wilfred Dolfsma

LONDON AND NEW YORK

First published 2015
by Routledge
2 Park Square, Milton Park, Abingdon, Oxon OX14 4RN

And by Routledge
711 Third Avenue, New York, NY 10017

Routledge is an imprint of the Taylor & Francis Group, an informa business

© 2015 H.L. Aalbers and W.A. Dolfsma

The right of Rick Aalbers and Wilfred Dolfsma to be identified as authors of this work has been asserted in accordance with sections 77 and 78 of the Copyright, Designs and Patents Act 1988.

All rights reserved. No part of this book may be reprinted or reproduced or utilised in any form or by any electronic, mechanical, or other means, now known or hereafter invented, including photocopying and recording, or in any information storage or retrieval system, without permission in writing from the publishers.

Trademark notice: Product or corporate names may be trademarks or registered trademarks, and are used only for identification and explanation without intent to infringe.

British Library Cataloguing in Publication Data
A catalogue record for this book is available from the British Library

Library of Congress Cataloging in Publication Data
Aalbers, Hendrik Leendert.
 Innovation networks : managing the networked organization / Hendrik Leendert Aalbers and Wilfred Dolfsma.
 pages cm
 Includes bibliographical references and index.
 1. Knowledge management. 2. Information networks. 3. Technological innovations--Management. 4. Diffusion of innovations--Management. 5. Information technology--Management. 6. Organizational behavior. 7. Strategic planning. I. Dolfsma, Wilfred. II. Title.
 HD30.2.A243 2015
 658.4'038--dc23
 2014043144

ISBN: 978-1-138-79697-3 (hbk)
ISBN: 978-1-138-79698-0 (pbk)
ISBN: 978-1-315-75752-0 (ebk)

Typeset in Bembo
by FiSH Books, Enfield, UK

Printed and bound in the United States of America by Publishers Graphics, LLC on sustainably sourced paper.

"An organization has no presence beyond that of the people who bring it to life."
Gareth Morgan (1993)

CONTENTS

Figures and tables   xi
Preface   xiii

1 Introduction: the networked organization   1

PART I
Networks and organization strategy   19

2 Diagnosing the organization   21

3 Innovation roles: internally and externally oriented brokerage   35

4 Intermezzo: cooperation for innovation at Siemens (case)   52

5 Rich ties: combining networks   63

6 Cross-ties for innovative teams   78

PART II
Networking interventions: rewiring the organization   99

7   Intervening to stimulate innovation   101

8   Innovation despite reorganization: rewiring the network   115

9   Methodological considerations for network analysis   125

10   Managing innovation in the networked organization: conclusions   142

Appendix: supporting notes to Intermezzo Case "cooperation for innovation at Siemens"   147
References   155
Index   171

ILLUSTRATIONS

Figures

1.1	An intra-organizational network	3
1.2	Two nodes and a connecting tie	9
1.3	Various ways to look at network relations – directed and valued ties	10
1.4a	High redundancy network	12
1.4b	Low redundancy network	12
1.5	Networks and Centrality	13
2.1a	The innovation network at Redrock at t=1 (organization level)	29
2.1b	Zooming in on the combined networks of 5 randomly selected employees from the innovation network at Redrock at t=1 and their individual centrality measures (Group level)	29
2.1c	Ego networks at the individual level (Individual "ego" level), Innovation Network	30
3.1	Communication roles	41
3.2	The formal (mandated) workflow network	44
3.3	The informal network	45
4.1	The formal (mandated) workflow network	59
4.2	The informal network	60
5.1	The informal, formal, and multiplex ties	70
5.2	The Innovation Network for Greenwood	71

6.1	The formal vs. informal elements in an organization	81
6.2	Horizontal and vertical cross-ties	82
6.3	Horizontal and vertical ties	87
6.4	The Innovation Network at Redrock	88
6.5	Horizontal ties' contribution to team performance	90
6.6	Vertical ties' contribution to team performance	92
7.1	Innovation network, before and after an intervention (t=1 and t=2)	111
9.1	Data scoping	128

Tables

2.1	Centrality calculations at the individual level (Individual "ego" level)	30
3.1	Number of individuals in the five different communication roles	46
3.2	When are which brokerage profiles most beneficial?	48
5.1	Descriptives – Frequency of tie types	72
7.1	Ties in the innovation network, t=1 and t=2	112
9.1	An overview of common organization network types	132
10.1	Key network take-aways for innovation	146

PREFACE

Intra-organizational networks – the topic for this book and its underlying research – has developed and matured for us over quite a few years. During that time – in which we both published and consulted on a plurality of aspects of the phenomenon – we grew our understanding of what organization network analysis is, and how applying a network-based view within an organization can be beneficial – enabling innovation to thrive. Organization network analysis is now on the verge of going mainstream, having developed on the fringes of the social sciences for a long time. Through this book we challenge the false idea that social networks are ephemeral, changing shape rapidly and therefore making them difficult to understand and hard to manage purposively. In the process we open up a variety of analytic and academic network techniques and insights to a managerial world.

What is attractive about organization network analysis is that it allows for very precise predictions about what behaviors and outcomes of behaviors can be expected, given the structure of the connections between actors observed. These insights and understandings complement the more interpretive insights that social sciences have offered, but sometimes challenge these as well. The precise predictions arise from the clear and elaborate tools that have been developed over time to tease out ever more detailed aspects of social network structures that might impact behavior and outcomes.

In addition to allowing for more concrete predictions, translating or providing further precision to existing theories, organization network analysis is in fact a theoretical approach in itself. If not from its inception, then now. Through its singular emphasis on the structure of interactions, rather than their content, a different emphasis emerges from what is emphasized in other theories in the social sciences.

Partly, what readers will find in this book is a reflection of the precision that organization network analysis provides to existing theories. In that respect, many readers will find it relatively easy to understand what we offer in these pages. In addition, humans are inherently social beings, and so will understand and welcome the insights to gauge their social environment better. Throughout the book we provide insights that we have ourselves added to the field. While based on rigorous academic research that we have conducted, we have aimed to offer insights drawing on this that are accessible to a broad audience. On occasion, however, we have chosen to provide clues of the rigorous research involved in organization network analysis. For those who would like to pursue such research themselves, and for those who would like to have a better understanding of the outcomes of such research, we also provide a chapter with methodological and other research considerations.

As with the topic of what we discuss here – how innovation involves collaboration among multiple individuals that is structured somehow – so with our work on this manuscript: we have hugely benefitted from a number of different people, in multiple ways, and we would like to acknowledge this gratefully.

Over the years, we have worked with Rene van der Eijk, Hans Hellendoorn, Otto Koppius, Roger Leenders, Salvatore Parise, Dave Rietveld, Sander Smit, Jasper de Valk, and Eoin Whelan to further our understanding on a variety of aspects of organization network analysis. We are thankful for the enjoyable and insightful conversations and debates on the fascinating topic of organization networks. We also owe gratitude to Deloitte Consulting and in particular to Pieter Hofman for supporting the early exploration of the various ideas that form the foundation of this book.

We have been fortunate to be able to discuss our insights and suggestions for further work in different settings. We have presented our work at the Organization Science Winter Conferences, DRUID conferences, the SMS special conference on micro foundations of the firm, EGOS colloquia, an International Conference on Innovation and Management (ICIM), an International Product Development Management Conference (IPDMC) conference, and of course multiple International Network for Social

Network Analysis (INSNA) Sunbelt conferences, and we have been very happy to debate and reflect with participants there.

We have also presented our work at seminars in different universities, including Bocconi, Copenhagen Business School, NUI Galway, University of Glasgow, Lund University, Montpellier Business School, and Rotterdam School of Management (RSM). We have learnt how best to convey our message about organization network analysis, but gained new insights as well, from teaching our students. And finally, this book gained much from responses by those that bring the organization to life, as we presented our insights to a variety of practitioners with whom we engaged throughout our studies.

1
INTRODUCTION
The networked organization

Organizations are highly dynamic social entities. They are collections of individuals that collaborate to produce something that none of the individuals could produce by themselves. This poses problems – but also offers opportunities – to management and employees alike in complex organizations. In these pages, we introduce a view of the organization as a set of overlapping networks that are vital to the development of a strategic competitive advantage of the organization: the *network based organization*.

The search for innovative potential within the organization evidently is not a new one. How to manage, measure, and profit from innovation has received considerable scholarly and managerial attention over past decades (c.f. Davila et al. 2012; Tushman and O'Reilly 2013; Li et al. 2013; Anderson et al. 2014; Lee et al. 2014). These same studies point out that making innovation work has proven to be anything but a straightforward task. Many organizations that depend on their capacity to be at the forefront of new product and service offerings in a diversity of markets and to a plurality of customers have encountered that innovative knowledge does not easily spread inside organizations. An organization that can improve the spread of knowledge internally will be more innovative (Bartlett and Ghoshal 2002; Aalbers et al. 2013).

To compete effectively, being able to innovate continuously is a must. We have found – based on our research and consulting experience – that being able to scale and intervene in organization networks sets apart trailblazing organizations from the rest in this respect. Innovation results from the combination and recombination of existing and newly developed pieces of knowledge. Knowing how to identify the critical resources that can serve as the foundation for new products and services to be brought to the market early on therefore is of high value to most growth aspiring companies. In an organization, having knowledge available and accessible at a moment's notice to the right people, ensures that it can be innovative, responding quickly to the highly dynamic environments it operates in. Intra-organization networks provide the social infrastructure to manage such exchange effectively.

The structure of social networks and the nature of the ties in them provide essential cues as to what the social interactions in an organization may be expected to deliver. To understand the inner workings of organization networks an understanding of the core notions of organization network analysis and the underlying network theory is essential. This chapter sets out to review these core notions as a point of departure for a fact-based approach to rendering insights into the networks within an organization, as brought to the fore in the remainder of the book.

It will have become evident by now that a powerful way in which to understand the social nature of organizations, and the leading way in which to analyze this, is by looking at the organization as a constellation of different networks. A network, in its essence, is the interconnections between nodes constituted by ties. Inside an organization, the people are the nodes who exchange (have ties) with others. The interconnected relations, the ties, show relationships or flows between the nodes. These simple notions easily combine into an intuitive picture of the social fabric of intra-organizational relations, an intra-organizational network (see Figure 1.1 for an example). Managers need to learn the language of networks to understand their organizations better. Others, including fellow scholars, would also benefit greatly from understanding organizations as a constellation of networks.

Here we introduce and discuss key terms drawn from the field of network analysis such as tie, node, tie strength, and centrality, and place these in an organizational context. We do so from an almost technical point of view, and also, importantly, from a strategic point of view, that takes innovation as a prime driver of strategic advantage, as well. We focus on what these terms mean in the context of business and zoom in on intra-organizational networks. Understanding network concepts and the avenues they offer for purposive intervention will augment organization performance and help management to steer towards innovation and sustainable competitive advantage.

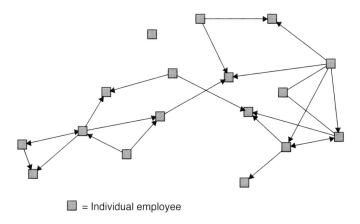

FIGURE 1.1 An intra-organizational network

Intra-organizational networks may be instrumental, affective, strongly mandated, or largely discretionary. Accordingly, various networks can be identified, from formal or workflow networks that are strongly mandated by management, to informal networks that may or may not form the shadow-vertebrae of an organization. What is actually exchanged is – or can be – very different in each of these networks. The knowledge that is actually exchanged in a network is highly context dependent as well. For an accounting firm, for instance, in one network, accounting data is exchanged, while in another network advice about how best to relate to customers is exchanged, in still other networks new ideas and knowledge that will contribute to innovations for an organization is exchanged. There are networks of individuals discussing last night's football or basketball game, as well as networks of those who smoke together during breaks. Each network has its own merits to those involved. These are all different networks, each with their own structure and dynamics over time. Individuals can be involved in more than one network at the same time. Any two individuals in an organization can then be connected with another individual in the organization in more than one way. Adapting a view of the organization as a set of diverse, and overlapping networks, provides an alternative path to more traditional modes to understand and enhance the strategic competitive advantage of the organization.

Approaching the organization as a constellation of networks offers an *intuitive* yet *thorough* insight into its functioning. Social networks are easily understood, since they are close to just about anybody's immediate understanding of family, friends, and society. Networks can be visualized readily thanks to the latest advancements in the field of Information Technology and network analytics, and can be monitored over time, offering an additional intuitive appeal. More fundamentally, based on a combination of sociological and mathematical (graph) *theory*, organization network analysis helps us to determine how the social influences individuals face affect their behavior in ways that are often not comprehensible to the individuals themself.

Organizations increasingly become (socially) complex, even when they are small. In large part this increased social complexity is due to the stronger emphasis by firms to continuously innovate. Firms have to innovate faster to adapt to changing market conditions: three-quarters of the 1,500 global senior executives taking part in the annual Boston Consulting Group Innovation Survey judge innovation to be crucial for their business growth strategy, identifying innovation as a top three priority for their company (Wagner et al. 2014). Despite the high priority and increased spending on innovation, 70 percent of these same executives rate the innovation capability of their firm as only average – and 13 percent indicates it even as weak (Wagner et al. 2014). A continuous flow of innovation can create sustainable organization competitiveness and growth, but is not achieved easily (Moran 2005).

Innovation is, however, a fundamentally social activity. While our insights pertain to other activities that are important to organizations, we therefore focus primarily on how organizational networks shape organizational innovativeness. In this book we thus offer insights on how the networks in an organization affect an important activity of organizations that determines their competitive position now and in the future: innovation. Our goal is to better understand how social networks affect the extent to which an organization is innovative, and offer insights into how innovation may be stimulated by leveraging these social networks in their organization.

Innovation in networks

Knowledge is the most valuable asset and an important source of competitive advantage for an organization (Grant 1996; Teece et al. 1997). Scholars have emphasized that effective transfer of knowledge between employees within an organization increases the organization's innovativeness and creativity (Davenport and Prusak 1998; Tushman 1977; Moorman and Miner 1998; Perry-Smith and Shalley 2003; Tsai 2001; Hansen et al. 2005).

In social networks, information is generated, dispersed, screened and enhanced (Campbell et al. 1986; Coleman 1990; Granovetter 1973). This can provide all kinds of benefits, to all kinds of players outside of a focal firm. Entrepreneurs who are properly networked are more likely to succeed. Venture capitalists rely heavily on their networks when making investment decisions. Even in rather mature markets, where changes are rare and incremental, and where one as an employee, business unit leader, or CEO are well aware of someone's individual circumstances and the circumstances of one's peers and competitors, one's network of contacts determines the extent to which one will be able to differentiate, and ultimately to be more successful than someone's competitor. If someone has friends in his network who are particularly innovative, this person is more likely to be innovative too. Such contacts are likely to span beyond that person's immediate, daily, social contacts. A network provides actors with access to valuable information well beyond what someone could process on its own (Burt 1997). Someone's network surrounding may actually act as additional discretionary processing capacity to one's own processing capability (Kijkuit and van den Ende 2010). While Information Technology may make much information accessible too, the sensitive, complex (partly because tacit) and valuable information is likely to be available through personal connections only. In order to make the best use of these connections, it is vital to be positioned in a network in a way that does not lead to information overload, while still allowing one to tap into a diversity of creative and original potential. The personal connections are important to screen and enhance information (Kijkuit and van den Ende 2010).

Simply aiming to have "more transfer of knowledge" is unlikely to be successful, however. Knowledge transfer is by no means self-evident and automatic (Szulanski 1996). Knowledge usually is spread throughout the organization and may not be available where it might best be put to use (Cross et al. 2001; Moorman and Miner 1998; Szulanski 2003). Many actors may be involved with the flow of knowledge in an organization, and even if they are willing to exchange it, they may not be aware of who might be in need of what knowledge they have. Knowledge proves to be one of the most difficult resources to manage. A clear understanding of the relevant networks in an organization, their structure and dynamics, is crucial.

Successfully organizing innovation processes is no mean feat. According to Joseph Schumpeter, innovation is the combination and recombination of knowledge. Combining or recombining existing knowledge can in itself constitute (the basis for) an innovation, or it can give rise to the development of new knowledge and insights based on the combination. The sources for these different pieces of knowledge usually are spread across the organization, and outside of it, among different individuals. As knowledge from different corners in organization coalesces, not only are the results that come from it more likely to be technically superior, but the acceptance of whatever emerges is enhanced. (re-) Combining knowledge inside an organization is a social endeavor. Since innovation processes are socially complex by nature, managing innovation entails managing social networks. Social network scholar Ron Burt has shown that, as knowledge comes together at an individual level, this individual and the organization in which they are employed is much more likely to be innovative.

> Perceiving the organization as a set of overlapping networks that are vital to the development of strategic competitive advantage of the organization is a crucial step towards successfully managing innovation networks in business.
>
> *(the Authors)*

Organizations that leverage their innovative potential in a more intelligent way, using insights from organization network analysis offered in this book, will be more likely to cut it in the highly competitive global business environment. Even in relatively small organizations, the social infrastructure easily becomes complex. For example, between n employees employed within a firm, $(n \star (n-1))/2$ potential connections may exist in any one network. For an organization that employs 50 employees, the maximum possible number of connections among them thus is already considerable, 1225 to be precise. A manager of an organization will only have a partial view of what the *social infrastructure* in that organization looks like. What is more, the view that the manager has, of the *social infrastructure,* may be biased as the knowledge the manager receives, may be tainted.

This book offers managers and students/readers a way to better understand the social activity in an organization, formal, informal, or other kind. We offer academically tested methods for an even more thorough understanding of the social side of organizations. We hope and expect that even a casual reader will be tempted to explore the academically more challenging parts in this book. For those who want a more thorough understanding of organization network analysis, the appendices to chapters, or the method chapter might offer additional insights. The take-a-ways that we suggest at the end of each substantive chapter are thus both academically founded and managerially relevant.

In a way, then, seeing organizations as constellations of social networks that are dynamic over time helps in uncovering the innovative potential of the organization. This potential intimately ties in with the degree to which innovative knowledge can flow within the organization. As knowledge flows in an organization through the pipes of the innovation network, previously separated knowledge is joined.

From such a social network perspective, a number of issues will stand out that other perspectives will not allow one to see. There are, for instance, a number of different barriers to knowledge transfer. Incompatibility between the type of network position one is in, versus, what one is expected to contribute. For instance, suggest how one responds to incentive structures one faces. If one is to contribute to an organization's innovativeness, but is ill-positioned in networks to do so, as suggested by Organizational Network Analysis, this is a potential barrier for knowledge transfer to you, and will prevent one from being innovative. Transfer of one kind of knowledge (rather than another) might also threaten existing (strategic) positions of individuals. An important barrier is the lack of information regarding the knowledge that is available somewhere in the organization, on individuals who possess this knowledge, and on how best to transfer this knowledge once it has been located (Hansen 1999; Szulanski 2003; Winter and Szulanski 2001; Hansen et al. 2005). If some knowledge available in an organization remains unused, its strategic position may be hurt. Knowledge transfer might be among individuals with (partly) incompatible frames of reference. In this book, we focus a variety of social (network) related reasons why knowledge might or might not flow readily in an organization. This book does not dwell much on psychological and personal issues, however, but focused on the *structures* of social networks.

Organization network analysis has a rich history in the organizational literature (Ballinger et al. 2011). Networks, the social structure made up by a set of individuals (actors) and a set of the dyadic ties that represent a specific kind of relationships between the individuals, provide a useful analytical lens to assess the social systems that constitute an organization as well as the patterns of interaction with her environment. Although this book does not focus explicitly on relations individuals have to outsiders, how an organization relates via its networks to suppliers, customers, competitors, regulators, and other stakeholders makes a big difference in terms of unlocking innovative potential to the organization.

8 Introduction to the networked organization

Network tactics deployed by an individual in an organization network can lead to the acquisition of power as they are capable of controlling the flow of information within (a part of) the organization (Brass and Krackhardt 2012). This may be beneficial to the organization, but can also lead to serious disruption of an organization's routines and innovative potential. Organizational Network Analysis also allows one to determine where an individual can do damage to innovation activities and knowledge transfer. Managing the flow of knowledge within the organization is a valuable skill to master, requiring an understanding not only of the formal, but also the informal activities within the organization. Any intervention by managers may change network structures stimulating the flow of knowledge to the benefit of the organization, even if not intended to do so. Managers should try to understand at all times what relevant social networks in their organization look like, how they can be improved upon, and how interventions impact them. All individuals in an organization, in actual fact, might want to do so.

Networks: Structure and position

How a network is configured determines the extent to which information flows freely or effectively within an organization. How a network affects an individual's performance in an organization may come from beyond this person's immediate contacts: it is important to know what all the networks look like if we are to have a proper understanding of the networks in an organization. The interplay between the different networks further affects outcomes for individuals, as well as outcomes for the organization. Management as well as employees can alter the structures of most networks that exist within organizations, enhancing, disrupting or disturbing the flow of information thus affecting the innovative potential of the organization.

In this book we suggest a number of different concrete and readily implementable tools to understand and alter social networks within the organization. Some are elaborated upon more than others. All lead back to a number of basic notions identified by network scholars over the past decades, now made readily available and placed in the specific context of an organization striving to unlock its un(der)-tapped innovation potential. Our book thus provides managers as well as students and scholars a method for studying social networks within an organization from a number of stances. This includes a set of descriptive, quantitative techniques for creating statistical and graphical models of people and their relations and interactions in an organization. The outcome is a better understanding of why some individuals and organizations are more innovative and perform better than others.

A basic understanding of social network concepts is needed for this, however – consider these the building blocks for an analysis of networks. Some network characteristics play at the individual and some at the organizational level.

Ties indicate that (during a defined time period), connected *nodes* have somehow interacted. This is shown in Figure 1.2. The nature of the interaction indicates the kind of network studied – one must make sure not to lump different kind of interactions (different networks) together. Nodes, often individuals (although they may also be projects, Business Units (BU's), organizations, regions, patent classes, or journals), and the relations (ties) between them, form the key building blocks of a network. Ties can indicate advice giving, formal collaboration, or the exchange of new innovative knowledge. Ties can also indicate a citation from one project, or academic journal, to another, the co-classification of patents in multiple patent classes to signify overlapping knowledge fields, or the acquisition made by an organization in one region of an organization in another. In the context of this book, nodes are most intuitively thought of as individuals, and ties as relations between them.

The more nodes that interact in a network, the larger the *network size*. One may consider a network in which a person operates as a resource which can be drawn upon. The *direction* of a tie provides an indication of how information flows in a network. Arrow heads indicate the direction of the knowledge flow between two nodes. When information flows from one actor the next, this is referred to as a directed tie. Advice relations, or relationships based on hierarchical directive are examples commonly found in an organization. If the information flow simultaneously takes place in both directions, this is referred to as a reciprocal relationship. Examples of these are specialist relations, where knowledge is exchanged to enhance the overall understanding of a technical problem, or the informal relations to keep each other posted of the latest company gossip or account developments. Within a network, the intensity of interaction between the different nodes may vary.

FIGURE 1.2 Two nodes and a connecting tie

10 Introduction to the networked organization

In case of close interaction between two nodes, rather than loose and intermittent interaction, one speaks of strong ties (rather than weak ties). *Tie strength* can thus differ between ties. Tie strength refers to the nature of a tie between two individuals (nodes in a network). Frequency of interaction can be a useful measurement of tie strength, but sometimes perceived closeness is a better one, depending of the kind of network investigated. Tie strength can also be indicated by the intensity of collaboration, the degree of trust among actors, as well as the awareness of the other's domain of knowledge. Tie strength is typically reflected by the thickness of the tie when visualizing network relations. The thicker the depiction of a tie, the stronger the tie strength between those interacting.

The relevance of taking the value of ties into account depends on the objective one has in mind when looking at an organization. In some cases one is only interested in the quantity and configuration of intra-organization relationships. For instance to study the relation of network structure on one's network position to assess individual conduct and anticipate individual performance. In this case one takes a *Structural Embeddedness* perspective of the organization. This perspective ignores organizational or individual characteristics. When, instead, one is interested in the quality and contents of intra-organizational relationships, this is referred to as taking a *Relational*

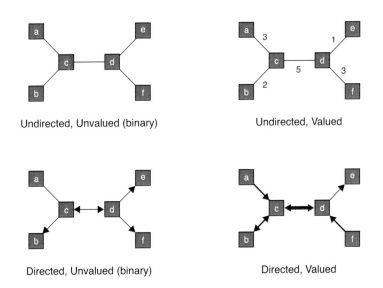

FIGURE 1.3 Various ways to look at network relations – directed and valued ties

Embeddedness of the organization. In this case, intra-organizational relationships are viewed as a source of competitive advantage, and intra-organizational relationships are perceived as a potential source of innovation. This book provides the tool and analytical angles to do take on both perspectives, depending on what one aims to achieve.

Illustrative of a relational embeddedness perspective of the organization, Granovetter (1973) has argued that a large network of ties, which necessarily consists of *"weak" ties* mostly since communication costs would otherwise be inhibitive, helps to obtain and disperse new information and to mobilize people and resources. Such a network will also allow an individual to more easily "get a job" (see Granovetter 1995). Granovetter (1982: 130) famously stresses the importance of weak ties:

> The importance of weak ties is asserted to be that they are disproportionately likely to be bridges as compared to strong ties, which should be underrepresented in that role. This does not preclude the possibility that most weak ties have no such function.

Too often an argument drawing on social network conceptions suggests that weak ties are favorable in all possible circumstances. *Strong ties* can be helpful as well, however, in particular when complex, tacit knowledge is to be exchanged, knowledge that requires an understanding of sources and background to be fully understood (Hansen 1999). When a tie is strong, the people involved will know each other better, are more likely to trust one another, which facilitates the exchange of tacit knowledge or knowledge that could make either of the parties involved vulnerable (Bouty 2000). In any case, it will be more likely that what the other communicates is properly understood.

Related to Granovetter's "strength of weak ties" argument is what Ronald Burt has suggested about people who connect two or more otherwise unconnected (sub-) networks (Burt 1992). Those suggestions do not focus on the quality and content of relations, but on the structural position individuals take on to broker information, evident of a *structural embeddedness* view of the organization. Such "structural holes", or "*brokers*", as he calls them, will be able to exert control over the information flow between these networks in each of which knowledge may be available that is relevant for the other. Burt (2004) has also shown that people thus placed are in a better position to develop new ideas themselves.

At the network, rather than the individual node level, these concepts help to understand the discussion on the extent to which networks (in an

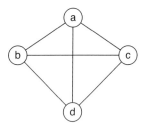
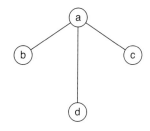

FIGURE 1.4a High redundancy network

FIGURE 1.4b Low redundancy network

organization) should actually exhibit *redundancy*. A network that is fully connected is one where all persons are tied to other persons: network density is at a maximum. Figure 1.4a shows this. Few networks are like this however, since maintaining these ties, even when they are weak, is costly. All knowledge being at the disposal of everybody at all times inhibits specialization, and increases the communication costs tremendously. On the other hand, in case of low redundancy in a network, knowledge lying with a single specialist is likely to go unnoticed, allows that specialist to extort rents, and makes the organization vulnerable when they leave. Coleman (1988) thus argues that there should be (some) redundancy in a network for it to work properly. In a low redundancy network some persons are better positioned to exchange information than others. This is shown in Figure 1.4b. In a high redundancy network each person has access to the same information sources – this gives no one the powerful position, but does imply h   communication costs to maintain the large number of ties.

Some parts in a network may interact more cl ly than other parts – high density (redundancy) can be local. In network rms this is referred to as *clustering*. This is a phenomenon much studie in economic geography: effects of organizations clustering regionally have been believed to be highly beneficial since at least Alfred Marshall (Van der Panne 2004). Such organizations may be more successful, for instance in terms of innovativeness (Giuliani and Bell 2005). Silicon Valley is an example of this (Saxenian 1994), one that many other regions and countries try to emulate. Co-location or frequent interaction in a network increases the chances of discovering and exploiting opportunities (Thornton and Flynn 2003; Burt 1992). At the same time, frequent interactions locally in the social infrastructure may, due to ensuing high communication costs, prevent people from interacting with others in a network that are positioned outside of the cluster. Knowledge and information that these others could well use might not reach them. This may be an unintended, and unappreciated, effect of clustering. Clusters or cliques in a network can also be used intentionally by actors to promote a specific purpose such as to be better able to mobilize politically against others in the network (organization).

As one of the most important measures in network research, *centrality* indicates the influence of an individual within a network with regard to other individuals within that network (Brass and Burkhardt 1992). Centrality provides an evaluation of the position of individuals within the network in terms of how close they are to the "center" of the action in a network. Centrality measures are about the influence or importance of individuals in a network, but some are only meaningful and measurable if one has an understanding of and information about the full network.

In Figure 1.5d, person "b" is highly dependent on person "c" to provide information, but person "c" can try to obtain information from "a" as well as "b". In Figure 1.5b, person "a", in the center of the star, can draw from

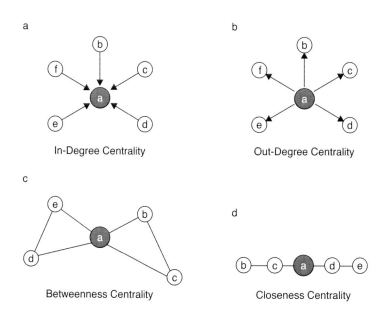

FIGURE 1.5a–d Networks and Centrality

alternative channels when the requested or sought after information is not provided by one contact in the network. If person "b" decides not to provide "a" with information, "a" can easily turn to person "c" or person "d". "b", however, does not have that option, which makes it likely that "b" will provide "a" with the information solicited in the first place.

Although the notion of centrality has several interpretations, in this book we look at centrality of an individual actor active in a network in terms of:

1 Degree centrality: counting the number of in- and/or outgoing ties from one's position within the network. This is an intuitive measure: more contacts means more actors one can draw on. Since contacts may not be reciprocal, it makes sense to distinguish in-degree from out-degree. The first says how many others have mentioned someone; the second says how many someone has mentioned as a contact. In academic studies the first tends to be preferred since it does not rely on self-reporting. In Figure 1.5d, all except "b" and "e" have 2 as their degree centrality. "b" and "e" are clearly not favorably positioned.
2 Closeness centrality assesses the "distance" of a person to all other persons in a network. Closeness centrality is an intuitive measure of how well (quick) one can actively spread information oneself or become aware of information, rumors and epidemics spread by others through the network (defined as the total number of steps of the actor to all the other reactors).
3 Betweenness centrality assesses the extent to which one is positioned on the shortest path between any pair of any two of one's colleagues in the network. Betweenness centrality is an intuitive measure of how one plays a role as go-between, broker, or structural hole. Betweenness centrality is likely to be the best measure of centrality if control over the interaction processes (information exchange) is important. In Figure 1.5c, person "a" is most central when measured in terms of betweenness centrality.

In addition to Degree, Betweenness and Closeness Centrality, Bonacich centrality is an intuitive measure of the status that someone has since it takes into account the number of onward ties someone an agent is connected with has. It bases one's centrality score on one's neighbors' centrality. A higher centrality of this variety indicates strong control over one's direct contacts, and the contacts of these contacts. This provides someone with a high Bonacich centrality with alternative sources of information and can allow one to strategically disseminate information.

Closeness, Betweenness and Bonacich power centrality are measures that make sense only in case one has larger networks to study, and they require that one has a view of the full network. Some of these centrality measures are sensitive to network size and cannot simply be compared between networks. What is more, centrality measures are based on a snapshot, a still of a particular network at a particular point in time. As networks can be dynamic and change over time, this means that reflection on network positioning benefits from multiple observation points in time. The stability, or robustness of a person's network position benchmarked against the positioning of others as the network evolves is an interesting indicator of network positioning at the individual level.

Centrality measures help to identify persons with a high information processing potential in a network. Highly central persons are highly involved in transferring information and are therefore highly aware of what is going on within the organization. Centrality is an indication of the degree to which individuals have access to resources including most distinctly knowledge (Hoang and Antoncic 2003). Centrality is therefore a useful concept when studying innovation in an organization (Ibarra and Andrews 1993). Interpreting centrality measures in isolation is undesirable however. The social embeddedness of the individual in the overall network helps to assess the degree to which an individual is (not) effective, or by contrast, may suffer from information overload due to a lack of alternative brokers.

The innovative organization: Network tactics

Exchange of information is essential for an organization to foster a climate of innovation in any company. An organization must create an appropriate social infrastructure as a means to this end. For a manager seeking to stimulate organization innovativeness, a view of the organization as a set of (partially) overlapping social networks, each with their own structure and idiosyncrasies, is indispensable. The exchange of knowledge and information is dependent on the structural characteristics found in the networks of an organization. The importance of this, if not clear by now, is best gauged by posing three questions:

1 How easy is it for an employee to reach out to others within the organization?
2 How difficult is it for an employee to tap into the most recent insights of that what is going on within the organization?

3 How much effort does it take for an employee to contact others to establish future collaboration?

Managers, but others in an organization as well, can use the insights from this work to tactically change the networks or network positions in any network that they are in. Network tactics play at both the individual and the network level, and suggested actions at each level may be conceived.

Network tactics at the individual level. Information exchanges take place through different networks constituting the organization's social infrastructure, and the individuals can determine the degree to which the flow of information is smooth or haphazard. While individual characteristics and motivations are important, what they can do is largely determined by where they are located. Actors who are favorably located can do more to facilitate/ hamper the flow of knowledge and information than those who are at the periphery, intentionally or unintentionally. Individuals who have many ties, whose contacts again have many ties, and who have ties with important others are in a good position.

Network tactics at the organization level. At the organization level, structural holes, those connections to disconnected sources of non-redundant information, drive innovative ideas (Brass and Krackhardt 2012; Burt 2004). Individuals might find themselves to be a structural hole in a network, but cannot (easily) seek to maneuver themselves into such a position.

Managers who are well-positioned like this are also likely to be more effective. Managers who are mindful of the importance of social connections in the different networks can seek to change the social infrastructure in order to facilitate exchange of knowledge and information in their organization (Brass and Krackhardt 2012). One can build a coalition in support of a purpose for an organization, and stimulate the flow of information and knowledge that comes along with that. Coalition building can take place in several different networks at the same time, with activities across various networks reinforcing the likelihood of reaching the goal of, for instance, the organization becoming more innovative.

Some exchange of information is important to the business but *difficult to manage*, such as exchange in the informal network of work related issues. Informal exchange can sometimes even hurt the organization, for instance when gossip undermines a manager's position or when it hurts progress of what management considers a key project or relevant strategic direction. However, managers may influence some networks more than others. Managers control who formally exchanges with whom, but to a far lesser extent who informally exchanges with whom. Management can formally set up permanent and transitory structures such as organizational governance models and day-to-day project teams. They can and should do so with a view of bringing together people with the right knowledge and skills. What we know is that people who are formally brought together to collaborate, are also likely to establish connections informally. Managers can in this and other ways, indirectly, influence network structures beyond their immediate control as well. In line with the renowned law of unintended consequences coined in the early twentieth century by sociologist Robert K. Merton, the outcomes may be positive, resulting in unforeseen benefits such as enhanced knowledge flow, but may also be negative, resulting in disruption of the social system. Management can also intervene in networks more directly, as will be discussed in some of the later chapters; for instance by relocating or dismissing individuals, but should be careful to consider what favorable network connections may get disrupted.

Main take-aways for Chapter 1

- Innovation results from the combination and recombination of existing and newly developed pieces of knowledge.
- Intra-organization networks provide the social infrastructure to manage such exchange effectively.
- Intra-organization networks are vital to the development of the strategic competitive advantage of the organization.
- Having knowledge available in an organization accessible at a moment's notice to the right people ensures that an organization can be innovative, responding quickly to the highly dynamic environments it operates in.
- Innovative knowledge does not easily spread inside organizations.
- An organization that can improve the spread of knowledge internally will be more innovative.
- Knowledge transfer within an organization should be studied from a network perspective, both at the organization and at the individual level.
- A social network is a set of nodes connected by ties (social relations) with specific content and objective.
- The structure of social networks and the nature of the ties in them provide essential cues as to what the social interactions in an organization may be expected to deliver.

- Organization Network Analysis (ONA) provides fact-based insights into the networks within an organization, which people actually collaborate (most), or fail to do so, and how knowledge spreads inside an organization.
- Employees, and managers in particular, can then better mobilize organization resources around strategic challenges, translating strategic goals into the day-to-day practice, in order to achieve sustainable business transformations, including, most prominently, innovation.

PART I
Networks and organization strategy

2
DIAGNOSING THE ORGANIZATION

A focus on the structural features of the networked communication in an organization is a key factor when management seeks to stimulate organizational innovation. Here, we provide insight into which communication patterns are most effective when organizations seek to become more competitive and innovative, or simply aspire to secure their current level of competitiveness and innovation. We identify characteristics in organizational networks that are beneficial for organizational innovation and provide directions for a proper network based diagnostics of the firm.

Networks in organizations

A prime type of exchange that takes place between people in an organization, is *the exchange of information and knowledge related to new ideas and innovations*. The relationships within an organization that facilitate this kind of exchange is called the *innovation network*. Exchanging such knowledge is key for organizations in today's competitive environment. In the end, no organization can expect to survive if it does not innovate.

Transfer of knowledge and information in an organization is not self-evident, even when everyone involved understands its importance and are motivated to contribute their part. For instance, the relative autonomy of divisions within a typical multi-unit organization structure frequently creates a lack of awareness about what is going on and what knowledge is available beyond an employee's own unit. This restrains knowledge-transfer and the recombination of knowledge and resources. Such recombination has been identified as crucially important as a means of creating innovation and competitive advantage in contemporary strategic management and entrepreneurship research. Leaders in an organization have the opportunity to intervene, directing the exchange of knowledge in ways that benefit the organization. Before management intervenes, a thorough diagnosis of the organization's networks is vital. This chapter elaborates on how to *deploy* some of the network concepts introduced previously to effectively diagnose an organization. Further chapters provide additional depth and breadth.

What we know about knowledge exchange in networks

The better connected an employee, the more they contribute to innovation and profitability. Those with diverse connections contribute more, while "linking pin" employees who connect otherwise disconnected groups do best of all – for themselves and for the organization. People who connect well vertically do better for themselves and their team, unless there are a large number of others in their environment who do so as well – thus making the team ineffective. Of course there can be too much communication taking place within an organization – when all are connected to everybody else, leading to high density in a network. The costs of communication may then outweigh the benefits. While in theory, at some point, the cost of communication may start to rise over and above the benefits, in practice one is unlikely to encounter this situation. No research, including our own, has provided evidence of this occurring in practice: most organizations do not exchange knowledge enough.

Organizations face various challenges in leveraging resources for innovation effectively. The following are exemplary questions for senior management:

- How do we identify how and where innovation in the organization happens currently?
- How can we harness individual knowledge to increase collective strength?
- How can we exploit the most connected employees to influence large groups?
- How do we select for senior influencing roles?
- How can we harness informal networks to support formal goals?
- Generally: How can we identify new opportunities (the innovative potential not currently, actively being tapped within our organization)?

For each of these questions, Organization Network Analysis (ONA) proves a valuable analytical device, helping to leverage the power of human networks in organizations. ONA offers many features. Properly taking a network based lens to the organization can help organization leaders amongst others, to:

- Identify different relevant networks in an organization, including the informal and the formal networks.
- Reveal cross-geographical, cross-business unit and cross-functional connectivity.
- Support increasingly flexible organization designs, e.g. Project based team structures.
- Unveil the information channels between groups and individuals.
- For a given organizational goal, identify the most influential employees.
- Reveal how and where things get done (or are blocked).
- Indicate personnel risks that require hedging strategies.

These attractive outcomes for managers from application of social network analytics, however, require an understanding of how to diagnose an organization network, how to interpret the outcomes of these diagnostics, and how to translate these outcomes into actionable follow-ups.

How to diagnose

Insight into how the different communication networks in an organization are structured can provide important clues as to why the organization is performing the way it is. Organizational network analysis (ONA) has developed into a systematic approach and set of techniques for studying the connections and resource flows between people, teams, departments and even whole organizations. Various applications of ONA have been developed to capture and mine organizational network data varying from the relatively straightforward collection of social connections via workshops or questionnaires, to more advanced Radio-frequency identification (RFID) tracking employees as they go about their daily job, and longitudinal email or mobile data mining.

To effectively diagnose the organization network for innovation, it is important to have a clear understanding of the actual innovation needs. Organizational network analysis cannot determine which employees need to exchange what knowledge with which others, and how. Managers and experts of the content of an innovative knowledge field must determine this among themselves. Organizational Network Analysis can then be a diagnostic tool that provides managers with insights into innovation blockages and opportunities. Often its impact is simple, for example in highlighting the difference between senior managers' "view of the organization" and the ground-level "reality".

Assessing the climate for innovation

Assessing the innovation climate of an organization is not a simple task. Innovation draws on a plurality of potential ingredients that can grow into new business concepts and subsequently to innovative competitive advantage for the future. Organization network analysis allows management to conduct what we could refer to as an "innovation engagement scan" (IES).

An IES provides objective insight to help understand the shape and workings of networks within an organization. It provides insight into a number of general network characteristics, such as the degree of connectedness within a network, the diversity of these connections (e.g. the ratio between inter and intra business unit connections), and can identify the main players responsible for the (formal or informal) flows of knowledge within an organization. Through IES, the (potential) knowledge hubs available within the organization relevant to innovation, and the patterns of actual collaboration between people on innovation can be uncovered, and put to better use. In helping management to determine the strengths of an employee, team and organization's engagement with innovation, an IES demonstrates not only who exchanges innovative knowledge with whom, but also what profiles the contributors to the innovation community of the organization, or business unit, display. With these insights, management can mobilize and strengthen existing social networks around specific innovation challenges in order to improve the innovative capacity of the company.

IES bases itself on recent research which shows that the combination of formal and informal relations, for instance, is an essential requirement before the exchange of actual innovative content will occur. Chapter 5 elaborates further on this. Potential inhibitors for or resistance to innovation can be identified as well, for instance. Innovation Engagement Scan contributes four key insights. It:

- Identifies ideation gaps.
- Exposes silos.

- Identifies leadership gaps.
- Identifies opportunities and risks for continued innovation involvement.

In one example, an IES helped to assess the organization's climate for innovation inside the marketing and sales department of a medium sized manufacturing company. The IES first laid out the organization's innovation relations. A clear divide between both divisions came to the fore, with a "No man's land" between departments. Exchange of new insights between both domains predominantly took place only via other departments that served as an intermediary, including the production department. The IES that identified this was subsequently used to inform a workshop with representatives of the marketing, sales and production departments to determine the rationale behind the lack of direct connections. This process involved posing a series of questions (see below). In this particular case, a recent reorganization had resulted in a number of key persons leaving the organization. As a result, innovation contacts were mainly indirect, brokered ones. It was not clear as to which route knowledge could still take, nor where there were gaps. IES provided quick insights into this, as well as possible solutions.

Discussing innovation network outcomes – a checklist of topics:

- *Community determination:* When we talk about "innovation," with whom are we talking and how frequently do we do this?
- *Subgroups:* We talk and mail especially among each other about innovation, and less with others. Why?
- *Linchpin:* Which departments are often (or: rarely) involved in communication on innovation and who are the key individuals behind this?
- *Geographical breakdown:* Do we interact or mail beyond the boundaries of the business unit on innovation related subjects/matters?
- *Reaction to an event:* Who tends to be most receptive to innovation discussions, or reaches out to the outside world, and who falls back on passivity, waiting for others to act first?
- Is innovative activity more "alive" after a particular event/intervention?

Simply by improving relations between the key departments, innovative capabilities developed and even flourished again. According to the management of both divisions, the IES helped to mobilize the right people for the right positions, bridging some of the former relations that used to be mediated via colleagues who have left the organization. The glaring lack of linchpin activity and a deteriorated degree of online and offline interaction beyond the boundaries of the business unit proved eye openers to management and employees. Geographical breakdown was reconsidered in light of the changed organization composition, with strong managerial emphasis on interaction beyond the boundaries of the business unit as a priority for new project design.

The outcomes of the IES address a more general point that needs to be resolved prior to effective idea generation. In day to day corporate life, management may not be aware of the knowledge potentially available to employees in an organization. Employees in turn may not know what knowledge others in the organization have that they need, or may need in return. As a result of this asymmetry, the exchange of innovative knowledge can be hampered even when all are willing to engage in it. Individuals from two units might connect in one way – for instance: informally – but may not exchange innovative knowledge even though they should according to management, and also willing to do so. Connections between units may be maintained by individuals who are not formally required to exchange knowledge relevant for innovation. Relying on informal or even third-party connections in this way may be problematic if such connections are important to the organization yet continue to be the important kind of connection. An informal contact cannot necessarily be expected to persist, or have an organization demand that they do so. When they disappear, this may go unnoticed. Certainly if such a situation where formal connections between individuals from specific departments that are supposed to be connected are actually disjointed were to persist, actively establishing a direct link is certainly to be considered. If the connection between units is considered crucial, the actual connecting should probably be done by relatively senior individuals who have a formal mandate. Even without the assessment that this company went though, an organization can and probably should regularly check its communication networks to see how knowledge is actually flowing and what can be done to further stimulate that. This may well be an open and joint activity organized on a theme where current business and/or innovation strategy aspires to particular growth. Transparency on the diagnostics objectives is not to be taken lightly. Not only as proper communication of strategic intent driving this managerial action enhances engagement on the shop floor, but also for methodological reasons. A positive connotation on the work floor with the objective of a network diagnostics exercise enhances response rate and accuracy of the outcomes of an IES.

Thus far, we have focused on situations in which simply highlighting and facilitating relations and communications was enough to spur innovative activity and outputs. Here, a moderate intervention can cause individuals to shift their focus of attention to others in their environment. This stimulates interaction, dissolving some of the typical knowledge barriers, such as lack of familiarity among individuals, distinct thought worlds, disparities in verbal skill, status differences, and physical distance (Dougherty 1992; Weick and Roberts 1992).

Another case involves a consultancy. The IES revealed that new hires could easily integrate into the organization. A remarkable finding as the integration of newly hired personnel is a notorious problem for most organizations. While the IES pointed to a strength of this organization in terms of welcoming newcomers to its ranks to carry out the consultancy's day-to-day activities, it also showed that connections maintained by newcomers hardly progressed beyond the projects they were directly engaged in. The connections that had actually formed were also not necessarily appropriate to transfer more complex or sensitive knowledge. There was a need for further focus, improving information sharing within, but also among projects and towards the business units newcomers were formally affiliated to. The scan, in combination with a workshop, resulted in the following new insights.

The company needed:

- *Weak ties,* connecting individuals from different levels, subunits and tenure beyond their immediate daily functional environment.
- *Mentors,* well-connected with mentees as well as with the business units newcomers formally were affiliated to.
- *Stronger project networks* to improve informal learning among project teams.
- *To open up the business unit networks the newcomers are affiliated with:* both formally and informally, with lower dependency on only a few tenured individuals brokering access to these networks.

What does good look like?

When diagnosing the organization, a common question among executives is: what is the best structure of relevant networks in my organization? In other words: what does *good* look like? If one were to know what network structure to strive for, intervening is easy.

It is important to realize that network data in itself merely forms the point of departure for an organizational intervention. Understanding the structure and development of networks, however appealing their graphical layout may be, is not enough. Networks and network positions themselves are rarely objectively assessable, but should be understood in view of the organization's goals and circumstances, including the employees it has and their particular expertise and characteristics. However, analyzing networks in an organization is an effective way to understand why it (under)performs. A network picture must usually be combined with specific contextual knowledge for various reasons that we will not discuss at length here, as absolute network benchmarks do not exist. What works well for one organization or even part of an organization, can be counter-productive for another. What network configuration works successfully in one set of circumstances, may not in other circumstances.

The outcomes of IES can, however, foster directed discussion about, for instance, the reasons for innovation not to take off as expected, or it can point out both the most (or least) innovation-conducive elements of the organization. As an example, the interpretation of the centrality scores can help management and the employees assess who the knowledge leaders are (i.e., those with high brokerage scores), or those most likely to influence and be influenced directly (i.e., those that hold a high degree centrality score).

Consider the example of Redrock, a financial service firm. Innovation activities at Redrock focus on innovative payment methods, receiving significant attention by corporate management. At the organization level (Figure 2.1a), (random) group level (Figure 2.1b), and individual (ego) level (Figure 2.1c) exemplary network structures are presented, suggestive of the possibility to examine closely part of the organization's innovation activity as needed. Table 2.1 provides us with the calculations of the different centrality measures outlined in Chapter 1. Employee 12 and especially 21 are prime examples of employees heavily engaged in innovative knowledge exchange processes. One might alternatively worry that these persons are at the brink of information overload. A more balanced spread of knowledge leaders and influencers may be required, certainly over time. Even if individuals are not willingly limiting or changing the information flow, management may be concerned that the organization's dependence on these individuals for an organization's innovation performance is too risky. A similar assessment can be made for other relevant types of networks depending on the needs of the organization.

Diagnosing the organization 29

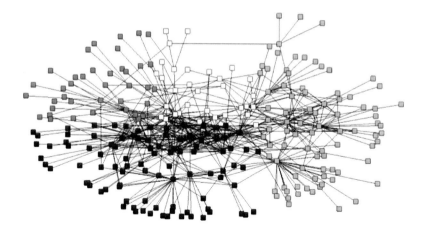

FIGURE 2.1a The innovation network at Redrock at t=1 (organization level)

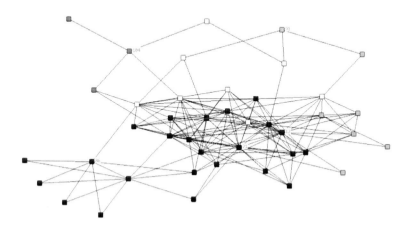

FIGURE 2.1b Zooming in on the combined networks of 5 randomly selected employees from the innovation network at Redrock at t=1 and their individual centrality measures (Group level)

30 Networks and organization strategy

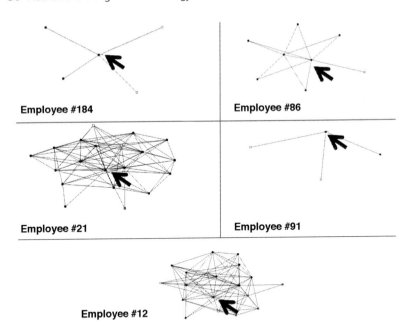

FIGURE 2.1c Ego networks at the individual level (Individual "ego" level), innovation network

TABLE 2.1 Centrality calculations at the individual level (Individual "ego" level)

Centrality/focal actor	Employee 184	Employee 86	Employee 21	Employee 12	Employee 91
Out-Degree (1 step)	0.012	0.016	0.039	0.02	0
In-Degree (1 step)	0.004	.012	0.059	0.055	0.012
Out-Bonachich Power	0.565	.57	40.908	12.274	0
In-Bonachich Power	1.686	.003	67.574	51.195	1.97
Out-Degree (2 steps)	0.071	0.016	0.184	0.086	0
In-Degree (2 step)	0.02	0.059	0.18	0.173	0.035
Betweenness Centrality	0.001	0.001	0.007	0.006	0
In-Closeness Centrality	0.530	0.533	0.531	0.531	0.536
Out-Closeness Centrality	8.863	0.397	10.801	10.417	0.391

Note: All are normalized scores, (that is one's score relative to all other employees' scores in the network).

A network based view to idea implementation

While the IES can help to improve the potential for idea generation in an organization, many organizations neglect to capitalize the opportunities ideation generates. How can organizations create a proper process to evaluate, select, and develop ideas without making the mistake of choosing low-hanging fruits and ignoring a promising idea with much more longer-term potential? From a network point of view: is low-hanging fruit supported by an appropriate social network structure? If not, they might appear to hang low but cannot be picked. An ideation process must thus be thoroughly embedded into the social structures of an organization.

To make an idea successful, it needs to be evaluated formally, but not just by looking at the content alone. Focusing on this only will make the (innovation) managers zoom in on:

- *Resources and Funding:* those higher in the organizational echelons decide upon access to resources and funds, convey messages and information, provide advice and help in case of difficulties;

A network based view of the organization, and innovation processes in it, stresses that an evaluation round should include more than only the recognition of idea quality, but should also involve an assessment of the social embeddedness of the idea. Therefore, there should also be a focus on:

- *Social Embeddedness:* employees from lower organizational layers have access to different types of resources, such as task-specific knowledge (Cooper and Kleinschmidt 1995; Han 1996).

New ideas should be supported by an appropriate team of individuals in an organization, with appropriate support from different echelons and corners. These individuals should themselves be well-connected, so they can, in support of an idea that they champion, tap into relevant knowledge sources and commandeer additional resources and support. Successful idea evaluation must pay attention to the possibility to socially embed the idea in such a way that it can tap into relevant networks within the organization that can further it along. As it is not easy to alter the social infrastructure of an organization, the assessment of the social infrastructure in support of an idea should be taken into consideration before deciding which idea to progress to the implementation phase.

Understanding how the organization's social infrastructure can be leveraged to allow an idea to grow entails a very different way for management to look at which ideas to develop, and which to abandon. Successful idea implementation requires *executive brokership* and wider organizational support. New ideas may upset the status quo, but having a sense of what relevant networks in an organization look like will allow management to localize resistance to a new idea. When implementing a new idea one will know where and how to tread carefully, and where support in the organization can be found.

Creating a climate that is supportive to innovations and to the possible changes these innovations may bring can benefit from directed network intervention. Network diagnostics can be applied to analyze the degree to which lower and upper echelon employees interact, support or hamper innovations, and if these patterns are likely to be effective. Our own research has shown, for instance, that for successful innovation teams, horizontal and particularly vertical cross-hierarchy ties are maintained by a small number of team members rather than scattered across a large number of team members.

Managerial intervention drawing on network diagnostics can use such insights to create a properly targeted "window of opportunity" for individuals in an organization to change the way in which they behave, potentially expanding their individual network horizon, adding new contacts, or revisiting the value and effectiveness of the individuals with whom they currently interact.

Rewiring and strengthening the organizational communication network optimizes the way *current employees* go about conducting their business. Welcoming newly hired individuals into the organization, and not only making them fit in, but also ensuring they make a fruitful and timely contribution in the near future is also much more effective when considering the existing social network structures that they will have to fit in with. Successful organizations make sure it is not just left to human resources or project managers to do this, but encourage all kinds of activities that will result in creating and further growing newcomers' networks. Management in general, and immediate supervisors too, can, for instance, assess the appropriateness of linking the newcomer to specific insiders. Depending on the skills, personality of, and goals for the newcomer, they can be connected to an externally oriented broker bringing in new knowledge from outside, for instance. But for a newcomer to be brought into a position to actively connect across unit boundaries might be disastrous since much tacit knowledge from different knowledge domains is needed for that. At the very least, what can be observed is that setting off with the expectation that connections relevant for innovative activity will simply emerge by themselves is likely to be a slow and frustrating process.

An intervention, light-touch or more, should also keep in mind the role within the organization that an employee holds or will hold in the near future. Individuals in particular functions may be expected to have a particular communication profile. For example, a unit director that only coordinates internally within a unit ultimately is ineffective as a leader. Network interventions redeploy individuals in particular functions, and if the new functions fits with a communication profile that they have, this is favorable to innovation. Both the unit director and junior employee can be induced to direct their attention to new relations relevant for the overall flow of innovative knowledge within the organization, in effect changing their communication profiles. What could be called a *communication director* creates connections, makes them more tangible and permanent, most likely enhancing the contribution of specific activities to the overall performance of the firm. Obviously, a communication director is versed in organizational network analysis and thus highly sensitive to the patterns of knowledge exchange in their organization, seeking to align these with the organization's strategic objectives.

Main take-aways for Chapter 2

- Studying organizations as networks is both complicated and a matter of common sense. The cliché: "I realize that people in an organization must connect, and as manager for this organization for many years I know who connects with whom!" is misleading: in an organization of any size beyond micro, a manager will not know what the different social networks look like that shape the performance of an organization.
- Interviews we conducted at board level at a large engineering company left no doubt about awareness at the top management echelon of the benefits to be reaped from structurally, yet efficiently, tapping into the social relations at the shop floor.
- These same interviews displayed a clear need with seasoned management to do so in an efficient way, without getting trapped into the common pitfalls that relate to micro management of the firm.
- People mostly have an idea of their immediate social environment, but their performance is also affected by two remote others who connect (or not).

- The organization's performance is shaped in large part by the *overall* structures of the social networks inside it.
- People's understanding of networks in their organization might, and even will, differ from their actual structure.
- Even when conscientiously collecting data, not having enough or the right kind of data can seriously jeopardize someone's understanding of the social fabric in an organization.
- Sampling and low response rates, damning for the research into social networks, can lead to the wrong diagnosis by managers of what happens now and how to intervene.

3
INNOVATION ROLES
Internally and externally oriented brokerage

What roles individuals in a networked organization adopt affects if, and how, strategic objectives such as efficiency or innovation are met. In a network based view of an organization, a crucial part of an IES (introduced in the previous chapter), innovation is fostered through a supportive climate. This view stresses the need for individuals to cross organizational (unit) boundaries. Effectively crossing boundaries in an organization and subsequently funneling new ideas and knowledge into the units where it can be put to good use is key. We present implementation recommendations which ensure that these new ideas disseminate successfully and efficiently from one part of an organization to other parts, crossing unit boundaries. Network brokers, the individuals that play an exceptionally active role in breaching unit boundaries, are individuals that an organization must nurture, but who are easily ignored. There are various kinds of network brokers, who are deemed to be the individual hubs of an organization's social network. Brokerage can be primarily (uniquely) internally or externally oriented. Depending on the strategic objectives of the organization, management can influence and steer how information flows by identifying and utilizing the most appropriate brokerage types. Depending on the overall characteristics of a business network, management might want to nourish, duplicate or even shield particular brokers to stimulate the organization's climate for innovation. The interplay between the various types of brokerage roles proves crucial to the innovative capacity of an organization.

Brokers

Social structures can help add value to the organization. The degree to which organizations are able to reap the benefits of the social capital represented by the networks, depends to a large extent to the degree to which resources can be accessed and mobilized through them (Adler and Kwon 2002). While the potential of leveraging an organization's networks to render innovative activity has been shown to be substantial (Aalbers et al. 2014; Ahuja 2000), the barriers faced within the organization to do so can be equally large. Organizational boundaries, such as organizational divides between different functional or operational domains, business units and departments, are hurdles for the free flow of knowledge inside the organization. This can be caused by: different Profit and Loss systems across business units, geographical dispersed activities, conflicting political sentiment or post-merger integration tensions. This is by no means an exhaustive list.

Obviously, organizational design and the definition of specialized business units is the leading reason for the existence of boundaries within an organization. Individuals who cross the boundaries found within the organization, have access to a broader range of information and receive that information earlier than others do. Brokers can facilitate transactions between others who lack access to or trust in others (Marsden 1982: 202). Consequently, brokers can combine and recombine existing knowledge, and have more potential for the creation of new knowledge, than others (Burt 2004; Kleinbaum 2012). Brokers, as knowledge leaders, are indispensable for any organization to unlock its innovation potential. The role of brokers should not be considered in isolation: brokers are the hubs of an organization's social networks, working *with* others who actively disseminate knowledge within the organization.

A focus on brokerage ensures that one considers who actually is involved in knowledge transfer and how (and in exactly which direction) knowledge flows. Contrary to centrality, group membership is taken into account (Granovetter 1973; Schulz 2003; Kleinbaum 2012). While unit boundaries in an organization exist for good reason, allowing for close collaboration and the development of specialized knowledge inside of them, managers must be aware that *directed* transfer of knowledge across boundaries can sometimes substantially improve an organization's performance. Theory and evidence advocates that organizations gain from an internal social structure rich in brokerage (e.g., Obstfeld 2005; Kleinbaum and Tushman 2007; Alcacer and Zhao 2012; Kleinbaum 2012). Identifying employee brokerage profiles by looking at an individual's communication profile, distinguishing helpful profiles from possibly problematic ones, offers managers the means to increase organization performance.

The degree to which brokerage occurs depends on more than just the structural characteristics of an organization's internal network. Research by David Obstfeld (2005) explored the behavioral orientation of employees towards connecting with others in their social environment. In addition to structural network characteristics, such as network density, and diverse social knowledge, a broker's strategic orientation is important: an active, strategic inclination to connect people in one's social network by introducing disconnected individuals or facilitating new forms of coordination between connected individuals is key (Obstfeld 2005: 102). This mindset is a "tertius iungens" orientation, and differs from an opportunistic orientation often characterizing brokers as playing people off against one another for their own benefit (a "tertius gaudens" orientation was originally introduced in the work of sociologist Georg Simmel). Individual characteristics of actors in a network are frequently overlooked yet may determine the ability to actually make innovative knowledge exchange come about within a firm (c.f. Aalbers et al. 2013). Still, whatever favorable personality characteristics an employee might have, if they are not well connected their contribution to knowledge transfer and subsequently to the innovation process can be limited (Aalbers et al. 2013). The remainder of this chapter therefore focuses on the positional characteristics of network brokers.

Internal versus *External* orientation

Network analysis commonly tends to ignore an important creation of management: unit boundaries. Bringing the right employees together into a single unit, working closely together can enhance performance and, at the same time, hold consequences for the knowledge that flows across an organization. Within a unit, specialized knowledge can be developed. As a result of such clustering of expertise one might expect most exchanges to take place within unit boundaries, facilitated by proximity as well as common procedures and practices. Knowledge transfer, the movement of useful knowledge or information between individuals, should also take place between business units (Appleyard 1996; Cummings 2004). Knowledge from other units will contribute to the knowledge developed within a unit to be more readily applicable in a new product or new services. Knowledge from outside will ensure that the knowledge developed inside a unit is more relevant, its development will be more broadly supported. In general, innovation activities are hard to carry out in isolation anyway.

Overall orientation of one's contacts – unit-internal, or rather oriented externally to individuals in other units – is known to affect employee performance (Gupta and Govindarajan 2000; Schulz 2003). Brokers who explore across unit boundaries within an organization secure access to outside information and are more open to innovation-related activities (e.g., Whelan et al. 2011). An employee who is externally oriented, in the formal or the informal workflow network, has greater potential to contribute to the innovative capacity of the organization (Perry-Smith and Shalley 2003). Employees who have an external orientation, for example, are more aware of existing knowledge sources within an organization that resides outside their own unit. Having access to knowledge not available in one's own unit makes externally oriented individuals more important compared to internally oriented ones in the earlier phases of the innovation trajectory (Hargadon 2002; Whelan et al. 2011).

Agency arguments make brokerage an important concept to understand in organization network dynamics (Ahuja et al. 2012). A broker can establish new relations or dissolve existing ones, depending on the benefits derived from the current brokering position. When seeking to improve one's network, one may either reduce one's own dependency on others, or, alternatively increase other people's dependence on oneself (Gulati et al. 2012).

However, the benefits of being externally oriented are hard to reap for an organization when knowledge cannot be put to use within one's own business unit. (Whelan et al. 2011). Knowledge brought into a unit from outside, needs to be absorbed, developed, and possibly transformed inside the unit (Ancona and Caldwell 1992a). For that to happen, there is a need within the unit for employees with a predominantly internal orientation. An employee capable of translating the typically diverse and heterogeneous knowledge acquired from across unit boundaries to the context specific needs and capabilities of the receiving business unit are a great asset to an organization (Burt 2004). Without internally oriented employees, innovative knowledge could never turn into successful innovations that actually make it to the customer in time. Employees that are mainly internally oriented can establish the necessary common base needed to integrate different knowledge sources (Ahuja 2000; Cohen and Levinthal 1990). Lack of common language and shared meaning might be detrimental to any potential innovative outcome, were it not without the help of these employees that broker based on a strong external orientation, transforming information so it can be understood by the receiving party (Dougherty 1992; Tortoriello and Krackhardt 2010).

While internal and external orientations are likely to have different functions for an organization, one actually may combine both to stimulate the transfer of new, innovative knowledge. When it comes to innovation, we suggest that externally oriented roles are likely to be most important (von Hippel 1994; Whelan et al. 2011). Due to the multiple networks that constitute the network based organization, an employee in point of fact may be internally oriented in one network while being externally oriented in another.

If and when this happens, benefits of multiplexity, where two individuals combine two different types of relations, allow for even richer and more accurate information exchange to occur, combining insights from one context to supplement, or simply validate, the other (Aalbers et al. 2014). Multiplex combination can be as simple as the maintenance of, for instance, a formal and an informal relations between two colleagues, not only working together on functional tasks, but also keeping each other informed on what happens within the firm that is of person relevance to either of them.

It is important to realize that a person's network orientation is dynamic by nature, reflecting individual preference, opportunity for change provided by the overall network structure, as well as resulting from management intervention (Aalbers et al. 2013; Ahuja et al. 2012).

Individual communication roles and profiles: Direction of knowledge flows

Kahn et al. (1964) were among the first to underscore the importance of "boundary positions" within an organization. They linked this notion to the maintenance of in-depth contacts of an employee with employees from other organizational units, or even outside the company. Employees may have externally oriented contacts, but they may mostly *receive* or *send out* information across unit boundaries. Only sending or only receiving knowledge across unit boundaries is likely to be sub-optimal and unsustainable in the longer term: there needs to be give as well as take (Ensign 2009; Dolfsma et al. 2009). Analyzing the direction of communication in the formal and the informal networks, as it crosses unit boundaries, generates additional insight into the antecedents for knowledge transfer. The concept of the network role(s) that an individual maintains is a useful one to capture this (Whelan et al. 2011; Fernandez and Gould 1994).

Several authors have categorized network roles by referring to an individuals' membership of a social group. Merton (1968) distinguished between the "local" and the "cosmopolitan". The local is mainly oriented towards his direct social environment, promoting social integration, while the cosmopolitan is more interested in the world outside his own community, stimulating social differentiation (Taube 2003). In an organizational setting, Allen (1971) focused on the technology gatekeeper. Boundary spanners acquire, translate, and disseminate external resources within the organization (Whelan et al. 2011). Individuals who carry out boundary spanning responsibilities gain status and influence through access to unique knowledge, but also experience significant role overload as a result of facing simultaneous and often conflicting pressures (Kahn et al. 1964; Katz and Kahn 1978; Marrone et al. 2007).

The identification of boundary spanners, or the distinction between locals and cosmopolitans, can be further elaborated upon in a way that helps to grow the understanding of the organizational antecedents relevant for transfer of new, innovative knowledge within the organization. In Figure 3.1 we present an exhaustive list of five different ways in which one may broker between two others.

Shapes of the nodes indicate unit membership

The focal individual is the individual top-center; the arrow this individual sends out (to the right, bottom) indicates if their orientation is primarily external or internal. Internally oriented individuals broker towards others in their own unit, even if the knowledge that they broker originates from outside, and externally oriented individuals broker towards others in other units. Coordinators and Gatekeepers are thus internally oriented, while Representative, Consultant and Liaison are externally oriented. Individuals can fulfill multiple such roles, even in a single network. The combination of roles of an individual can be considered their overall communication profile. One may assume that individuals should have a particular kind of communication profile depending on their position in an organization. Directors of a unit should coordinate within their unit, but should also represent their unit externally. An innovation manager will mostly liaise, certainly in the early phases of a new project, and may even take on the role of consultant, but should strive for others to take over their knowledge transfer activities. A person's expected communication profile can be compared with his actual profile by analyzing network data.

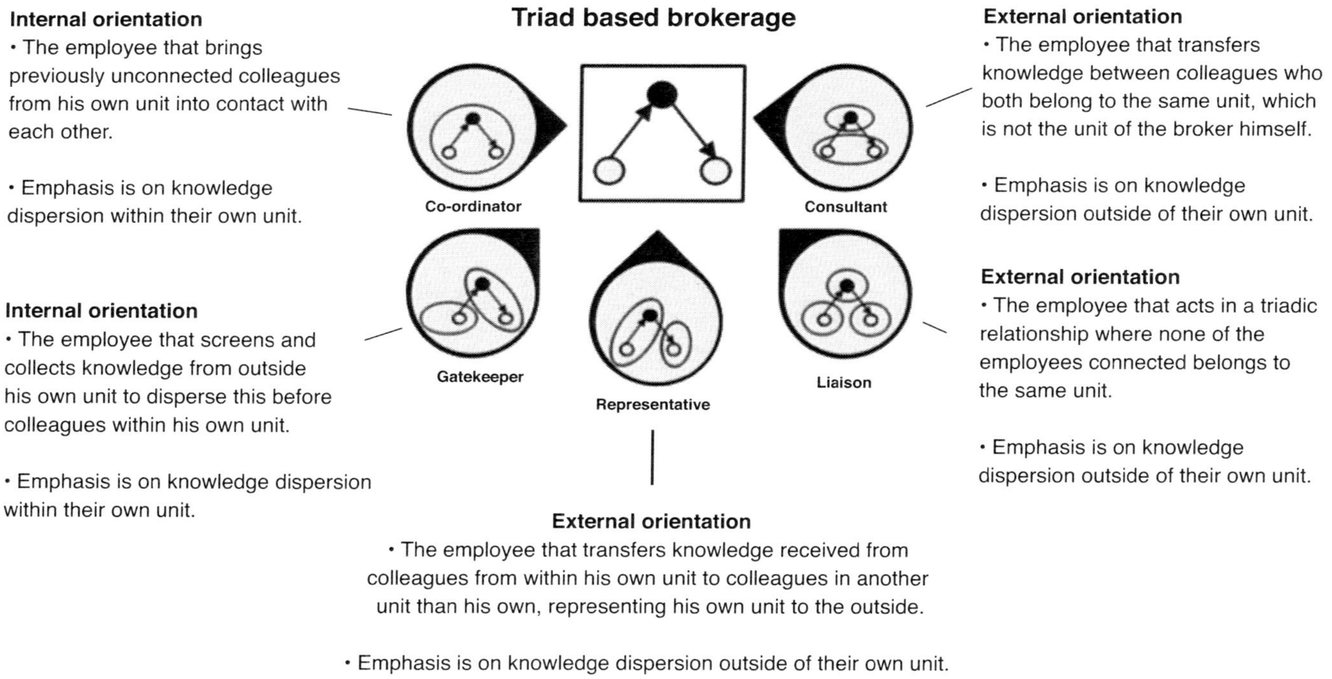

Triad based brokerage

Internal orientation
- The employee that brings previously unconnected colleagues from his own unit into contact with each other.

- Emphasis is on knowledge dispersion within their own unit.

Co-ordinator

External orientation
- The employee that transfers knowledge between colleagues who both belong to the same unit, which is not the unit of the broker himself.

- Emphasis is on knowledge dispersion outside of their own unit.

Consultant

Internal orientation
- The employee that screens and collects knowledge from outside his own unit to disperse this before colleagues within his own unit.

- Emphasis is on knowledge dispersion within his own unit.

Gatekeeper

Representative

External orientation
- The employee that acts in a triadic relationship where none of the employees connected belongs to the same unit.

- Emphasis is on knowledge dispersion outside of their own unit.

Liaison

External orientation
- The employee that transfers knowledge received from colleagues from within his own unit to colleagues in another unit than his own, representing his own unit to the outside.

- Emphasis is on knowledge dispersion outside of their own unit.

FIGURE 3.1 Communication roles

An employee brokering between any two other employees in a network, as a structural whole, may be more or less involved in either of the two sub-networks. The coordinator may be able to purposefully mobilize their network to advance an idea brought in from outside by a broker. Managers must look beyond the connections that particular individuals or employees have to more indirect connections that may be leveraged. What *network horizon* (indirect connections) do employees have?

Externally oriented brokers

Externally oriented brokers are integral to an organization's ideation and innovation process (Whelan et al. 2011). They act as the R&D unit's antennae, tuned in to emerging scientific and technological developments from around the globe. But while externally oriented brokers are very well connected to knowledge sources outside the company they tend to possess very few strong connections internally (Whelan et al. 2010). Without an effective internal distribution network, their contributions to an open innovation strategy are limited. Idea-exploring abilities, vital to the company's innovation objectives, may largely get wasted, when externally oriented brokers may lack an effective distribution channel to disperse the idea within the organization. This is exactly why network brokers should not be viewed upon in isolation. Brokers are as effective as their connections to others. The prime counterpart of the externally oriented broker is the internally oriented broker.

Internally oriented brokers

Internally oriented brokers are the hub of the company's social network, the go-to people of the organization (Parise et al. 2006). Much of their expertise lies in knowing who is doing what, and whom may benefit from what additional input or insights. When made aware of an opportunity for innovation, internally oriented brokers not only know who in the company is best equipped to exploit that idea but also possess the social capital needed to rapidly deploy the network to meet that particular challenge. Importing outside ideas is only part of the innovation challenge because new ideas will always encounter internal barriers. Leveraging the internal network to actually adopt those ideas is crucial. An external orientation is important for any innovation to emerge. Yet these key brokers simply cannot do it alone in the context of todays' modern, knowledge intensive and increasingly complex organizations. Internally oriented brokers play an important role in an organization's ideation and innovation process to push fruitful ideas identified by an externally oriented broker forward. The importance of network brokers to the innovation process has long been recognized (e.g., Allen and Cohen 1969; Obstfeld 2005; Aalbers and Dolfsma 2008; Lee 2010). The effectiveness of internally oriented brokers comes from the legitimacy of what they are seen to be doing in an organization. Our recent work on network brokerage has led to the identification of an additional, crucial brokerage role to be taken into account: the executive broker.

Executive broker

The role of executive broker highlights the relevance of hierarchical relations in the context of network brokerage. The jump from idea to sustained innovation depends to a large extend on the executive broker. The executive broker has a position relatively high up the hierarchy and is able to act as an ambassador to a new idea by increasing visibility of the idea to an upper echelon audience. An executive broker is also capable of allocating resources to a project, important in particular in the early phases of the innovation trajectory. This allows an internally oriented employee on the shop floor to actually dedicate their time and energy to connect the idea to the right person, or department, within the organization at the operational and tactical level. Strategic support is vital in particular in an environment where many ideas are being brought to the fore simultaneously, as is the case in many companies that run ideation on the basis of crowdsourcing or more traditional idea initiation procedures.

Empirical evidence on brokerage: Siemens

At a subsidiary of Siemens – a leading engineering and electronics company – the conditions for close cooperation between business units of its Dutch quarters were not optimal. The Intermezzo chapter, following this chapter, provides further background to this observation. To move towards more cooperation between units would be no mean feat for management to establish. The challenge faced by the organization at the time of our present study is captured well by the following quote:

> inter-divisional cooperation requires a radical change. The way people think needs to be changed: people need to operate more from their motivation of being an entrepreneur.

Figures 3.2 and 3.3 present two of the networks that were unearthed. Node shapes represent units. Note that peripheral nodes of individuals who did not have further contact are removed from the figures for clarity. To help understand the shape and workings of Siemens' innovation network an IES was conducted, providing objective data for further assessment. What analysis of this data showed, and what was not immediately clear from qualitative analysis, was that individuals who have an external orientation, actually passing on knowledge to others outside of their own unit, are more innovation-active. Notably however, this is only true for external orientation in the formal network.[1]

In addition, the IES provided the following results: The majority of employees are not externally oriented, neither in the formal nor in the informal network, focusing instead on colleagues within their own unit. This is understandable, on the one hand, as such focus allows for specialization among individuals with a similar area of expertise, and prevents communication costs from spiralling upwards. A limited number of individuals within Siemens bear the brunt of the inter-unit transfer of knowledge. In the formal network, with a size of 110, there are 17 coordinators, 9 gatekeepers, 19 representatives, 5 consultants and 9 liaisons. Even in a single network, individuals can of course have more than one communication role. In the informal workflow network, with a size of 87, there are 14 coordinators, 5 gatekeepers, 12 representatives, 1 consultant and 5 liaisons. Most employees, in both the emergent informal as well as in the formal mandated workflow

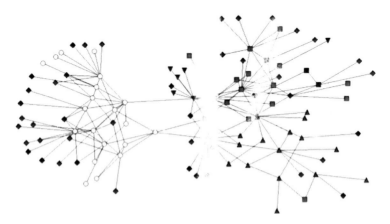

FIGURE 3.2 The formal (mandated) workflow network at Siemens Netherlands (N_{total}=110)

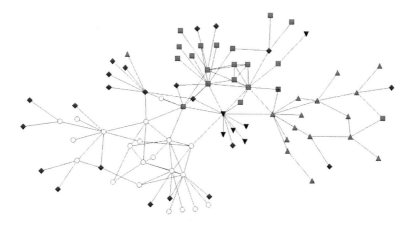

FIGURE 3.3 The informal network at Siemens Netherlands ($N_{total}=87$)

network thus start or halt the flow of information or knowledge – they do not pass on knowledge that they receive from others. When analyzing the number of unique brokerage positions individuals hold in each of the networks in Siemens the following pattern emerges (see Table 3.1). When the vast majority of employees are primarily focused on knowledge dispersion within one business unit organization, innovation suffers relative to an environment where externally oriented knowledge brokers are more active. Table 3.1 shows that, as expected, brokerage activity is lower in the informal network than in the formal network, since informal contacts are most likely to exist in the first place between employees who are in physically closer proximity.

Therefore, there is a strong dependence on a few employees for brokerage in general, and for exchange across unit boundaries in particular. Some units appear to be especially vulnerable, however, since they have fewer individuals with an external orientation than others.

This is a realization that can only emerge when one sees the complete network configuration in Figures 3.2 and 3.3. The unit pictured on the left, for instance, (the cluster of white circled nodes) is especially vulnerable to disruptions in the communication pattern with only two prime brokers maintaining ties to the other business units, aside from a number of bilateral relations with individual specialists from outside the unit. These specialists in turn are evidently not well connected, formally nor informally within their own units, as indicated by the scattering of nodes at the very left of Figure 3.2 as well as Figure 3.3.

TABLE 3.1 Number of individuals in the five different communication roles

Network:	Coordinator (i)	Gatekeeper (i)	Representative (e)	Consultant (e)	Liaison (e)
Formal	105	95	106	8	25
Informal	74	27	44	2	16
Innovation	60	35	57	3	24

Note: Total number of individuals in each role, in each network – individuals can have more than one role at the same time, in the same network; (i) internally orientated, and (e) externally oriented brokerage.

Obviously, managers should seek to enhance awareness of each other's expertise (Cross and Cummings 2004) as a precondition of being able to collaborate and transfer knowledge across unit boundaries. Simply, indiscriminately increasing the number of interactions and exchanges can raise the costs of communication, more than it raises communication benefits. A social network perspective focusing on employees' external orientation and also communication roles allows management to be specific about which employees to encourage to establish connections and transfer knowledge across boundaries with which specific others.

Communication profiles can differ between employees depending on their position and tenure. More senior positions tend to have more contacts to others in different units (Stevenson and Gilly 1991; Carroll and Teo 1996). More junior employees and those who have been hired more recently are likely to be more internally oriented and have matching communication roles. A deviation of someone's communication profile compared to what one may expect might be an antecedent for someone's performance in the organization.

Brokerage: connecting, exploring, sponsoring

Management can steer information flows constitutive of innovation notably by identifying and utilizing brokerage types. Brokers are the knowledge leaders within the organization. Depending on the overall characteristics of a business network, management might want to nourish, duplicate or even shield particular brokers to stimulate the organization's climate for innovation.

The following table summarizes why and how the interplay between the various types of brokerage roles, internally oriented or externally oriented in nature, proves crucial to the development of sustained innovative capacity for an organization. We find that connecting brokers with a strong external orientation must be connected with brokers capable of disseminating inputs within the organization, having a strong internal orientation (Whelan et al. 2011).

> **TEDx – innovation connecting brokers**
>
> An interesting example of the leveraging of these three types of brokers is provided by TEDx. The innovation conferences of TEDx are places where externally oriented individuals are sure to come into contact with radically new ideas and innovations. These events are organized with the sole purpose of exposing externally oriented individuals from a variety of companies to a plurality of new ideas and innovations that are likely to be new dots on their innovation radar. To enhance the chances of success, executive sponsors of leading organizations are invited to get involved in the innovation program of TEDx. The most promising ideas that are presented and explored during these events are then adopted for further screening and exploration by leading companies that provided a senior and executive level Executive broker, paired up with an experienced internally oriented broker from the same company.
>
> Executive brokerage in many instances means that organizations respond more quickly to innovations. At an individual level, externally oriented individuals and internally oriented individuals, who are responsible for the first part of the legwork, now see the fruits of their work in the form of recognition of their role and enhanced chances of realization of the projects they initiated. Placed in formal idea advisory boards, or operating casually and informally across the corridor of the organization, executive brokers provide support and the necessary resources. This means that the executive broker is more than the financial patriarch of an idea, but acts as a knowledge broker that smoothes out the political landscape and commits resources. By lending support, sponsors extend legitimacy engendering attention among relevant stakeholders, inside the firm (decision makers) and outside the firm (future core customers).

48 Networks and organization strategy

TABLE 3.2 When are which brokerage profiles most beneficial?

	External orientation – exploring	*Internal orientation – connecting*
Expertise	• Identify useful info and ideas outside. • Deep knowledge of particular (technical) field. • Strong analytical skills. • Strong online social media skills.	• The knowledge and expertise to connect conceptual ideas with internal skills and expertise. • Extensive knowledge and understanding on how to match new insights with existing knowledge base. • Skilled in translating and communicating external knowledge to internal colleagues. • Influential – ability to mobilize people.
Characteristics	• Wide network outside the organization. • An advanced degree in a specialized technological field. • Intrinsic interest to stay on top of latest technological developments. • Typically active through externally oriented social media to stay informed.	• Wide network within the organization. • Derives pleasure from helping others. • Has a good reputation and enjoys respect among colleagues. • Mostly active on internal social media to stay connected.
How to direct	• Provide with the time to explore the outside world. • Encourage them to attend remote (network) events. • Train in the effective use of social media. • Introduce to an experienced internally oriented broker. • Use a network scan to analyze and optimize the network. • Enroll in talent management programs and encourage scouting activity.	• Encourage to build and maintain internal network. • Encourage cross-functional project participation. • Link to community of externally oriented brokers. • Use a network scan to assess internal networks and identify any possible holes. • Enroll in talent management programs and encourage their internally oriented brokerage role.

Hierarchical brokerage – executive sponsorship

Expertise	• Decision making based on limited information. • Realizing public support. • Allocation of critical resources. • Recognizing talent.
Characteristics	• Decision maker. • Is open to new ideas, and encourages creativity. • Supports and protects brokerage at the lower echelons in the organization (shop floor). • Increases visibility and acts as an ambassador for new developments. • Stretches the need for the organization to innovate.
How to involve	• Involve from idea initiation phase. • Connect to the outside world, for example emphasizing flagship role through partnerships with other organizations. • Independent position as a senior innovation manager within the organization, backed by TMT. • Link to knowledge institutions such as universities and think tanks.

Network brokerage – conclusions

This chapter addressed the way in which organizations manage to leverage their internal networks to absorb external knowledge to enrich it, then to interpret and translate it into innovative and feasible solutions.

Brokerage of knowledge across unit boundaries can be undertaken by just a few individuals in an organization, making an organization vulnerable to disruption. Brokerage is, but possibly should not be, restricted to senior staff. Actual and potential employees who can broker can be identified by adopting a network perspective. Organizations that prove successful at this invested in a good balance between both the way externally oriented brokers, internally oriented brokers and the executive sponsor, broker to stimulate innovation. The interplay between these three brokers is decisive for the innovative capacity of the adaptive organization. Just absorbing promising ideas from outside the organization's boundaries does not suffice for an organization to be truly geared to innovation. Promising ideas will not develop into innovations, unless they are supported by networks of the right employees – employees that have the experience, knowledge and influence to nurture and implement them. Although with the advent of crowdsourcing and the Internet, the process for the discovery of new ideas has clearly changed, the critical role of the idea broker remains unchanged. Management needs to balance between the various brokerage profiles identified, the externally oriented broker, the internally oriented individual and the board level executive broker as agents of innovation.

Managers need to create conditions that foster network structures that drive enhanced performance, spurred by innovation (Balkundi et al. 2007; Varela et al. 2012). Viewing an organization as consisting of a number of different networks, and adopting a network perspective offers managers the possibility of analyzing their organization and defining targets (Aalbers et al. 2014). We suggest four organizational antecedents for the transfer of new innovative knowledge at the individual employee level, from a networks perspective: the *position* of the employee in both formal and informal networks, his overall internal versus external *orientation* in these networks, the employee's particular *communication profile* is derived from the combination of the latter two aspects, and lastly, attention to the *interplay* between the different communication profiles identified to broker innovation within the organization.

We know that within organization knowledge transfer crossing organization-internal boundaries is conducive to organization innovation.

Research, however, has thus far mostly focused on the centrality of the actors, mostly in the informal network (Burt 1992; Tsai 2001). The formal contacts between employees in an organization, and the way both of these types of connections are impacted by unit boundaries that compartmentalize an organization is, surprisingly, largely ignored (Reagans and McEvily 2003). Individuals in a network are often members of exogenously-defined business units – recognizing this helps us to understand what is going on inside an organization. This focus has its limitations. This chapter looks at innovation and focuses in particular on the direction of communication and knowledge transfer in an organization, showing that when unit boundaries are crossed, individual innovation increases. Managers can stimulate individuals to be more oriented towards employees in other units as sources for and recipients of useful knowledge. External orientation in the formal (but not the informal) network in particular contributes to innovation, which is fortunately where managers are best able to intervene in. In more detail, five different communication roles can be distinguished; roles which employees can adopt, in each network. Employees indeed can combine different communication roles, and organization innovation can benefit from it. More externally oriented employees, brokering knowledge within their organization across unit boundaries combining different communication roles will ensure that individuals can access the information or knowledge they need to *solve* actual *needs* of the organization – either within their own unit or in another unit.

Based on our observations and empirical analysis for each of the organizational antecedents we discussed, there are a number of suggestions to be made concerning what will likely stimulate innovative activity within an organization. Management can consider and seek to alter the internal or external orientation of individuals and even of units, and alter individual communication profiles, in both the informal network but particularly in the formal network. The findings presented above in particular provide a first indication that the brokerage concept that includes the *direction* of information flows is valuable for an organization when considering how to reorient communication patterns in their organization to support the innovative activities. Since managers are best able to change the formal structures in their organization, the finding that employees' position and orientation in the formal network stimulates their innovation activities more than those in the informal network is a vindication of their role.

Main take-aways for Chapter 3

- Brokerage is essential to innovation, as brokers have access to a broader range of information and receive that information earlier than others do.
- An organization that identifies and utilizes its network brokers is more innovative, as it uses what knowledge it has better, recombines knowledge into innovation more, and independently from these also creates more new knowledge.
- Intra-organizational brokerage benefits both the broker and the organization.
- Leveraging brokers requires the leveraging of a broker's social environment.
- Brokerage is not an individual action; management should approach effective brokerage as a team effort.
- Direction of brokerage matters, at both the individual and the team or unit levels, and needs to be balanced over time if some are not to lose touch with the organization.

Note

1 A Mann-Whitney test shows that external orientation in the formal, but not in the informal, network makes employees more innovation active (for the formal network: Mann-Whitney $U = 6.50$, $p = 0.016$, effect size $r = 0.632$; for the informal network: Mann-Whitney $U = 13.50$, $p = 0.279$, effect size $r = 0.30$).

4

INTERMEZZO

Cooperation for innovation at Siemens (case)

This case presents the problems and issues of communication in relation to innovation, as a dominant strategic objective. The case ends with a set of questions that can be used either by the individual reader or in a class setting and also serves for us as point of reference. Teaching notes are included as an appendix to this book.

Introduction

With more than 400 thousand employees in over 150 countries, Siemens is a world leader in electrical engineering and electronics. According to the business press, it is the world's largest, and Europe's strongest conglomerate. Its global network consists of 130 business units in six business segments: Information and Communications, Automation and Control, Power, Transportation, Medical, and Lighting. Each unit is responsible for its own worldwide operations, supported by regional units around the globe.

It is this decentralized structure that gives each business unit the flexibility to make its own decisions and build strong relationships with its customers. At the same time, cooperation among business units and regions is critical for enabling Siemens to provide comprehensive, customer-focused products, solutions, and services for the global market at a competitive price.

Siemens' worldwide vision at the time of observation was illustrated by the corporate campaign "Get a bit more, Siemens," under which the company was then operating. The campaign is strongly related to the products and services Siemens offers to their customers. Siemens strategy, at the time of research, was to offer total solutions to its customers, thereby fully responding to the customers' wants and needs. Creativity and experience provide the essential means to gain a strong competitive position. Siemens believes in a strong customer oriented approach, which means that the company is actively thinking about technical and societal issues together with their customers.

It is therefore not surprising that innovation forms the cornerstone in the success of the company. Siemens considers itself to be a worldwide innovation network that is formed by people who employ their knowledge in the fields of electrical engineering and electronics to the advantage of their customers.

Siemens files some 19 patents worldwide every day; a testimony to Siemens's innovative power. In total Siemens holds more than 45 thousand patents worldwide, ranking the company as the number two patent holder in Europe as well as in the United States. As a result of this, the concern derives three quarters of its total turnover with products that are less than five years on the market. Worldwide, this comes down to an annual expenditure of EUR 5.8 billion on R&D and the employment of 53 thousand full time researchers.

Siemens (SNL)

Although Siemens NL is not conducting much actual basic research through their Dutch subsidiary, SNL is very active in the development stage of new products and services. Like the parent company, the Dutch Siemens subsidiary is active on virtually all fields of electrical engineering and electronics. Its headquarters are located in The Hague, the Netherlands. SNL offers innovative technologies, custom made solutions and services for the market segments: energy, tele- and data communication, industry, healthcare, education, security and traffic, and transportation.

The situation at SNL

Reflecting the general structure of the Siemens mother company, Siemens NL is organized according to a matrix structure. This means that divisions have to report to the board of directors of Siemens as well as to the "Bereichen" (or fields/areas) of the German mother company. These areas can be regarded as the overarching business units of the German headquarters, located in Munich, Germany.

By positioning itself strategically in the market Siemens aims to increase profitability and expanded market share. The areas' main purpose is to increase global sales of their specific product-market combination. All divisions (or Business Units) at SNL are therefore accountable to two parties – to the Bereiche in Germany and to the board of the Dutch daughter company – and must take into account the local as well as the overarching strategic goals. As both parties have their own agenda this will sometimes lead to conflicting goals for the divisions to deal with. For the operations in the Netherlands however, the characteristics of the divisional structure can be regarded as dominant.

The management of Siemens – at the time of writing – operates its corporate strategy based on eight corporate themes. One of these themes is defined as "transport and mobility," covering a broad spectrum varying from the construction of specific trams to the transportation of goods. As the number of traffic jams is still increasing and competition within the European Union has increased due to the new legislation, especially the professional transportation sector has a large interest in smarter transportation solutions – be it by land, air or sea. It is for this reason that the management of Siemens wants to increase the company's focus on professional transportation as a part of the overall "mobility" theme.

Siemens targets the "transportation" sector dominantly by activities undertaken by its divisions BU 1 (Transport and Distribution) and BU 3 (Traffic and Safety). Regarding the transportation sector BU 1 mainly focuses on logistic solutions for harbours and airports, warehousing management systems and tracking and tracing. BU 3 mainly offers products and services such as traffic signalling systems, telematica (a combination of telecommunication solutions and information technology) and transportation systems (such as high speed trains, urban railways, and harbour logistics). Solutions are offered to governmental institutions as well as to business customers. Intermittently, other divisions are involved as well, among which are BU 2 and BU 4. BU 2 (Information and Communications) offers ICT solutions varying from specific SAP solutions to IT infrastructure maintenance. BU 4 (Mobile Communications) mainly offers communication related products such as mobile phones to the consumer market, as well as and in combination with solutions regarding broadband access and optical networks. However, for the business market BU 4 is also experimenting with tracking and tracing solutions and freight scanners it could offer in a more logistical context.

Although the first two divisions (BU 1 and BU 3) are active in the field of transportation, mutual efforts to join forces in approaching current and prospective customers have until so far never been really successful. There even seems to be a certain form of competition among the employees of the different divisions. The board of directors of Siemens has decided that the different divisions need to cooperate more to offer "integrated solutions" to their customers.

Striving for total solutions and an integrated market approach means that not only divisions with a current track record in the "transportation" sector need to cooperate more. Also divisions like BU 4 and BU 2 are therefore expected to engage more frequently in joint efforts between divisions. The conditions for closer cooperation between divisions however are not optimal. Incentive structures are heavily geared towards the performance of the individual divisions as well as towards short-term performance and profits. And although there exist several initiatives to facilitate cross-border knowledge sharing between divisions (such as Sharenet: an intra-organization software network aimed at sharing the latest knowledge between employees worldwide, innovation lunches etc.) the general feeling among employees and management is that there is a serious lack of insight into who knows what within the company.

The lack of knowledge about each other's activities makes it possible that a Siemens representative of BU 3 visits an organization to offer Siemens' view on the customer problems while at the same time an BU 2 representative is visiting the same office talking about the same problem, without both representatives (or divisions) knowing this of each other. The negative impact is obvious: besides the inefficiency of deploying Siemens sales force in this way, such activities also undermine the company's goals of offering total solutions to its customers. The situation is described strikingly by a BU 4 employee:

> it is often more easy to collaborate with another organization than with my colleagues in another division.

Bridging the boundaries

The problem of bridging the internal borders between divisions and between individuals forms one of the main concerns of the innovation management (IM) department of Siemens, headed by Hans, a seasoned change manager with a keen interest and knowledge of engineering technology. Operating as a staff department under the direct supervision of the board of directors, the IM department's main goal is to be a catalyst for innovation within Siemens.

The activities of the IM department include internal consulting, organizing "theme lunches" and other activities to improve the entrepreneurial spirit within the company. Besides this, the department monitors, assesses and incubates recent innovation proposals and projects and aims at improving the cooperation between divisions from an innovation point of view. Thereby, the IM department reports directly to the board of directors.

Ratified in strategic directions of Siemens, the company aims to annually achieve 30 percent of its annual turnover from inter-divisional projects. This in effect is quite a challenge, Hans realizes, as this percentage does not top 5 percent at the moment. The IM department needs to take action. Hans realized that reaching the company's goal of offering total solutions in which several divisions would have to be involved would be highly dependent on the way individual employees interacted across divisional boarders. To gain more insight into the actual flows of knowledge across divisional boundaries within the organization, the IM department, Hans decided to chart the communication flows within the Siemens organization, starting with the topic of "Transportation and Communication".

Interviews to shed some light on the employees' perspectives were conducted to gain more insight into the actual nature and size of the problem. Employees were generally found to lack (sufficient) insight into each other's activities and knowledge. Furthermore, employees perceived a misfit between the reward structure on the one hand, and requirement to increase interdivisional cooperation and activities on the other hand. They perceived this as a serious problem. Several employees working in the sales department of several divisions mentioned that to meet their current sales targets, inter-divisional cooperation was the last thing on their mind. One employee even stated that their supervisor reprimanded them for participating in an inter-divisional meeting that was organized bottom up by colleagues from another division. Several employees from different divisions seemed to share a common short-term vision when it came down to product offerings by Siemens. As one employee put it:

> We should not fool ourselves; all we do is 'moving boxes'...we should not regard ourselves so much as an innovative organization. Innovation is what happens in Munich, we are good at selling what they come up with.

The following statements illustrate this point:

> The current cost-structure limits the possibility to cooperate between divisions. Besides this, there is a lack of knowledge who knows what within the organization… A solution would be not charging each other internally for offered services, but working against 'actual costs' without the mark-up.
>
> (Employee BU 1)
>
> There is insufficient knowledge about the possibilities other parts of Siemens can offer us. We lack a clear structure and the dynamics for inter-divisional cooperation.
>
> (Employee BU 2)
>
> There are two problems, first of all we (colleagues at BU 3) have often no idea what other divisions can mean for us. Secondly, the cost/benefit structure is not effective. Often internal costs are higher than the external costs. Besides that, there exists some kind of competition between the divisions.
>
> (Employee BU 3)
>
> There exists a considerable amount of misunderstanding about each other's competences, unfamiliarity and incomprehension about each other's goals. Furthermore there are too many impediments for a total profit view where the loss of one division will be compensated by the loss of the other.
>
> (Employee BU 2)
>
> It is time for an overall 'stock-taking' of who knows what within this organization.
>
> (Employee BU 1)
>
> People in key positions should be granted more responsibility and trust instead of constantly having to justify their daily contribution regarding hours, turnover and so on. These people however need to be able however to carry this responsibility.
>
> (Employee BU 3)

Hans took the results from his study very seriously. If SNL was to be able to offer total solutions to their customers, the way individual employees interacted across divisional boarders would need to change. Up to that moment management had mainly emphasized in its incentive structure that bonuses would be earned for reaching goals set by every division separately. Communication between individuals across divisional boundaries appeared to be at least as important to achieve as inter-divisional cooperation. The following statement of a BU 4 employee, illustrates the awareness that the attitude of the individual employee is key when striving for organization-wide innovation:

> inter-divisional cooperation requires a radical change. The way people think needs to be changed: people need to operate more from their motivation of being an entrepreneur.

The networks

To further investigate the issue of inter-divisional communication, Hans decided to look at communication networks into more detail. He was interested in the communication profiles of divisions, and individuals within divisions, and realized that communication could take both a formal and the informal route. Both can contribute to the communication necessary to cooperate across divisional boundaries and develop the innovative products and services that SNL needs to develop. Based on Hans' earlier consultancy experience, these networks were described as:

- *Formal network:* the communication network that employees use to exchange information, documents, schedules and other resources to get their job done. It is based on communication patterns derived from formal procedures and company manuals and is reflected in the organization chart.
- *Informal network:* also called the "grapevine", is the communication network that is used to discuss what is going on within the organization and who is doing what next to the formal circuit. It is based on social relations based on *friendship and opportunity*.

As a picture can be worth more than a thousand words, Hans decided to visualize both these networks. Visualizing an organization's networks can indeed reveal some basic features about the nature of the company's knowledge flows. Based on a number of interviews followed up by a questionnaire, the IM department was able to visualize the actual communication patterns between and within the several divisions involved. The sociograms in figures 4.1 and 4.2 visualize the core of these networks:

The node shape indicates divisional membership, with employees depicted as "circles" in both sociograms representing the BU 1 division. The cluster of "squares" being employees active in the BU 2 division and the cluster of upward pointing "triangles" indicating the division BU 3. Furthermore, the small number of downward pointing "triangles" in both sociograms illustrate the peripheral position of BU 4 employees in the overall communication structures regarding the transportation theme.

The first impression

The visualization gave Hans a first impression of how open or closed the networks actually were. Looking at both networks it appeared that divisional membership formed an important reason for employees to cluster together. Although not totally unexpected this observation still came as something of a shock.

Clustering

When looking at the formal network, Hans interpreted the clustering of employees with other members of their own division, as a sign that there

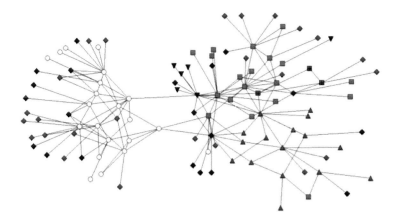

FIGURE 4.1 The formal (mandated) workflow network ($N_{total}=110$)

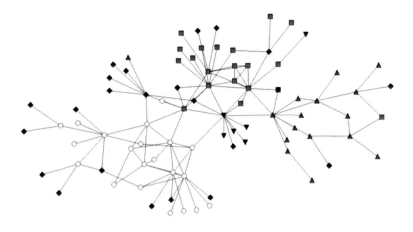

FIGURE 4.2 The informal network ($N_{total}=87$)

existed a lack of formal procedures aimed at integrating these divisions according to the company's purpose of offering total solutions to their customers. Looking at the formal network it appeared that there were only two people who made sure BU 1 did not completely lose touch with the rest of the organization. And among these two was Hans himself. Also, the relations between members of the different divisions at the right of the network did not seem overwhelmingly present. The picture of the formal network basically showed what Hans had already been predicting. The formal level of inter-divisional cooperation was not optimal.

Hans took a closer look at the informal network to decipher to what degree this situation was due to a lack of formalisation of inter-divisional cooperation or whether there was also a large cultural element in the tendency of employees to stick to their own division. Although the distinction between the different divisions was still apparent, Hans concluded that inter-divisional communication was slightly more common in the informal network. Surprisingly however, the division BU 2 appeared to be represented by only one employee. An interesting finding was that this same employee, in contrast with his relations in the formal network did discuss with a colleague representing the division BU 2. The same reasoning was applicable to a key BU 1 employee who was actually receiving information through the informal network about what was going on in the BU 3 division, while in the formal network he lacked any kind of contact with other divisions. Hans realized that inter-divisional relations were based more in the informal network than the formal network. Hans figured that these relations could be used to improve the formal network.

Reciprocity

The clustering of employees affiliated with either of the divisions in both networks was apparent. However, the way individuals were acting within these networks was not. Hans therefore took a closer look at the individual communication patterns of the different employees making up the network. By looking at the direction of communication and the kind of people an individual employee communicated with, Hans managed to group the employees into different communication roles. There was a lot of one-directional communication going on; in other words, reciprocal communication was rare. The direction of the arrows in the networks showed that employees who supplied other employees with information, in most of the cases did not directly receive information in return. Employees tended to gather knowledge for the benefit of their own division, but were not inclined to share information and knowledge that other divisions might be able to use. This proved to be the case for both the formal as well as the informal network.

Hans perceived himself as being aware of what is going on in the organization, and when he looked at his own position within the network he was not surprised to find that the relations with people from within most of the divisions of Siemens made him into a central figure. What was surprising was the fact that especially in the formal network the lack of other relations between BU 1 and the other divisions made his presence, simply put, indispensable. Without his presence the BU 1 division would be largely disconnected from the rest of the organization, and in the informal network this was also the case for the relationship between BU 1 and BU 4.

From an innovation point of view he realized this situation was based on a trade-off. Being the linchpin between several divisions was making sure that at least all activity and developments within the different divisions would not stay unnoticed to the IM department. But because being the head of the IM department involved other activities as well, and its role would have to be that of a catalyst, there was a risk that capacity restraints would lead to a situation where novel ideas and developments would not be acted upon in time. A sub-optimal situation for innovation to thrive could well result.

Hans realized this was a tricky situation, particularly after he noticed that besides himself, there were only a few people in the formal network who were fulfilling such a crucial role: be it within a division or between divisions. What would happen if these people were to retire, or, even worse, if they would suddenly transfer to competitors? There could be no back-up plan in place and valuable information would be lost.

These observations required Hans to act. Continuing with the existing communication practice would seriously hamper reaching the overall goal of gaining 30 percent of Siemens' Netherlands revenues from inter-divisional projects, such that the corporate strategic plan demanded. The question for Hans was, of course: how to change the communication practices at Siemens, preferably as soon as possible.

# 5
RICH TIES
Combining networks*

In particular when any two persons combine an informal and a formal dimension to the connection that they have, how likely are they to contribute to the organization's performance? Formal networks may be more important for innovation than informal ones; and informal ones, in themselves, may engender trust, but their combination in what is generally called a multiplex tie is even more beneficial. We refer to ties involving such multiplex interactions as "rich ties"; they are fertile grounds for in depth innovative knowledge exchange. The combination of various network dimensions is key to further develop an organization's innovation network. Once this is established, in this chapter, we draw out lessons to establish rich ties and make them work.

Rich ties in organizations

When the same two people connect in separate networks, they exchange with each other in different ways. The overlap between these networks is important, and is referred to as "multiplexity". We find that when people exchange formally as well as informally, in particular, they are also much more likely to be involved in innovation. Employees that maintain relations in separate networks at the same time also step up their involvement in innovation when management stimulates such activities, and are also more likely to maintain their involvement even after a major reorganization. We know that connections in organizations are mostly formal – there are many more purely formal connections than purely informal ones. Formal connections often are a basis on which informal connections can start developing. This is in sharp contrast to what many pundits, scholars, managers and laymen believe, that informal contacts lead the way for innovation. Fortunately, for managers, they do not. We submit that this is fortunate, as managers can intervene in the formal network of their organization much more readily than in the informal one. Emphasizing the informal network, at the expense of the formal network, negates the role of management. Emphasizing the informal in an organization can easily lead managers to not even start thinking about interventions that can have an immediate effect on organization performance. We emphasize that managers have ample opportunities to intervene.

Individuals in the organization can relate to each other in a number of different ways. When several dimensions of interaction between individuals overlap, these individuals have a multiplex tie (Burt 1983; Robins and Pattison 2006; Brass and Krackhardt 2012) and hence "quite different networks exist simultaneously within the same organization" (Lincoln and Miller 1979:182; Robins and Pattison 2006; Smith-Doerr and Powell 2005). Conceptually and empirically by identifying formal and informal networks in an organization, we determine whether these networks, separately, *as well as* in "multiplex" combination where they exactly overlap, explaining innovative knowledge transfer.

An organization's formal and informal networks are each well-recognized, distinct patterns of social relations in the literature (Rizova 2007; Gulati and Puranam 2009), recognized as antecedents for the transfer of innovative knowledge (Obstfeld 2005). Most network studies emphasize the importance of informal ties for effective knowledge transfer (Borgatti and Foster 2003; Hansen 1999; Krackhardt and Hanson 1993; Reagans and McEvily 2003; Rizova 2007). Although some older studies point to the importance of formal ties potentially contributing to knowledge flows in organizations (e.g., Darr et al. 1995; Nonaka 1994), formal networks have rarely been investigated in detail recently and even when they have been, they are often equated with the organization chart (Cross and Prusak 2002; Foss et al. 2010; Krackhardt and Hanson 1993; see Hansen et al. 2005 for an exception). However, formal and informal networks have, so far, not *both* been included in a single study that seeks to explain transfer of new, innovative knowledge in an organization. Attention to the informal ties dominates the larger research agenda as well (Foss et al. 2010; Gulati and Puranam 2009).

The formal and the informal networks each have their own distinct role in stimulating the transfer of new, innovative knowledge. At the dyadic level, we find, as we explain below, the shape of the innovation network is determined by the formal and the informal networks. When individuals connect both formally and informally at the same time, forming multiplex relations that constitute a network in itself, this gives rise to qualitatively different interactions (Burt 1984; Smith-Doerr et al. 2004; Brass 2012). However, multiplexity, the extent to which two actors are linked together by more than one relationship has been largely overlooked in studies that apply a network approach to intra-organizational settings (Agneessens and Skvoretz 2012; Brass, 2012; Grosser et al. 2012). The few studies that have included multiplexity, have not however focussed on innovation; and empirically indicate that benefits for individuals and organizations derive from entertaining rich ties. These benefits include increased intimacy of relationships and increased levels of trust (Minor 1983; Soda and Zaheer 2012), greater temporal stability of relationships (Minor 1983; Rogers and Kincaid 1981; Ibarra 1995), and reduced uncertainty at the individual level (Albrecht and Ropp 1984).

Innovative knowledge transfer in organizations, and rich ties

A central insight from the network approach to knowledge transfer in an organization is the observation that relations between individuals within an organization play a significant role in knowledge transfer (Allen 1977). While many different kinds of relations can be distinguished, a broadly accepted focus in the management literature is on formal networks of organizationally mandated relations on the one hand, and informal networks of emergent relations on the other hand (Allen and Cohen 1969; Allen 1977; Ibarra 1993; Gulati and Puranam 2009). It can be argued that these two networks are the primary ways in which people interact within an organization (Blau and Schoenherr 1971; Blau and Scott 1962; Simon 1976). Arguably, involvement in these networks would also, make the transfer of innovative knowledge in an organization more likely (Krackhardt and Hanson 1993).

Studies mostly tend to investigate one particular type of tie or network only (Hansen et al. (2005) are an exception), and in many cases the role that informal ties play in effective knowledge transfer is emphasized (e.g. Granovetter 1973; Hansen 1999; Powell et al. 1996; Reagans and McEvily 2003). Formal workflow networks have received much less research attention, and when they have, they are equated with the organization chart and found to be of marginal influence for knowledge transfer (Krackhardt and Hanson 1993; Cross and Prusak 2002). However, it would be neglectful and unfortunate to conceive of workflow networks as broader than just the organizational chart. Decisions to assign employees to divisions, work units, teams or projects all determine a person's place in the formal network of mandated contacts in an organization as well. Positions and connections in the formal network affect knowledge transfer as well, but have not always been considered in studies taking a narrow view of formal structures. Individuals in certain positions in the workflow network can be expected to impact the patterns in which knowledge flows in an organization (Allen and Cohen 1969; Aalbers et al. 2014; Stevenson and Gilly 1991). Despite assertions that different ties or networks can be conducive for different purposes or in different circumstances, a comparison between the different networks, for instance to determine which one contributes to knowledge transfer best, has rarely been undertaken to date (see Hansen et al. (2005) for a recent exception).

This chapter shows that formal relations contribute at least as much to knowledge transfer in an organization as informal ones. This is a vindication for the role of formal structures for knowledge transfer in the organization. After a first study to qualitatively compare the separate contributions of formal and informal networks to knowledge transfer (Gulati and Puranam 2009), we now provide a comprehensive, quantitative comparison. Rather than merely analyzing how the formal and the informal networks contribute to knowledge transfer separately, this chapter highlights their combined, multiplex contribution to the transfer of innovative knowledge. Combined informal and formal ties, forming a multiplexity network, turn out to be *rich* ties since they stimulate knowledge transfer more than ties in the formal-only and informal-only networks. Analysis of the organization networks of two purposefully different organizations (Cross and Cummings 2004; Levin and Cross 2004) are used to outline the concept of rich ties as "engines" behind innovative activity within the organization.

The formal network

Formal relations have been a historical focus of research among management scholars and sociologists (Aiken and Hage 1968; Blau and Schoenherr 1971), albeit without a strong emphasis on the transfer of innovative knowledge. Research on formal structures – "the planned structure for an organization" (Simon 1976: 147) – focuses on relations as stipulated by corporate management, most prominently in the organizational chart

(Kilduff and Brass 2001). Foss (2007) has argued that when knowledge processes and innovative knowledge transfer are discussed, formal organization are seldom, if ever, integrated into the analysis or are even neglected. Since the review by Damanpour (1991), the formal organization has not been the subject of much research in the field of innovation studies. Earlier research in management, however, mostly focused on the structural, mandated contacts in an organization as they followed from a formal employment contract and work protocols. The work of Mintzberg (1980) is an example of this. Formal networks have often been equated with the organization chart and were believed to indicate who reports to whom. In more recent decades, academic attention has become more of a transient, informal organizational phenomena (Krackhardt and Hanson 1993; Cross and Prusak 2002; Foss et al. 2010). Some scholars argued that formal relations or networks hamper creativity and demotivate individuals (Krackhardt and Hanson 1993; Robertson and Swan 2003). Others have indicated that formal networks reduce the autonomy of individuals involved in complex, non-routine activities (Tsai 2002). Formal networks have been claimed to reduce the flexibility of an organization to adapt to new circumstances and challenges.

In line with earlier network studies (Mehra et al. 2001; Brass and Burkhardt 1992; Gulati and Puranam 2009) we define the formal relations, which together form the formal network, as the prescribed roles and linkages between roles stipulated in job descriptions and reporting relationships. Formal structures are not limited to the organizational chart and include quasi-structures such as committees, task forces, teams, and dotted-line relationships that are formally mandated by the organization as well (Schoonhoven and Jellinek 1990; Ibarra 1993; Soda and Zaheer 2012). Even though the relationships in these quasi-structures can be more temporary than relationships represented by the organizational chart, they are mandated by the organization and part of the execution of daily operations in the organization (Adler and Borys 1996).

Formal structures, including quasi-structures, are relatively transparent. They allocate responsibility, and may thus prevent conflict and reduce ambiguity (Adler and Borys 1996). When an organization grows in size, a formal structure is required to stay in control and allow for specialization of tasks and knowledge (Adler and Borys 1996; Blau and Schoenherr 1971). The location of expertise is more easily determined and obtaining resources, e.g. for innovation, may only be possible by formal mandate. Thus, the formal structure dictates to a large degree who interacts with whom (Damanpour

1991; Gulati and Puranam 2009) and it is this formal interaction that could thus provide a foundation for innovation by the organization. As employees start to exchange simple, routine knowledge, this builds shared understanding, as well as absorptive capacity and competence trust at the dyadic level (Gabarro 1990; Lane and Lubatkin 1998), which can subsequently facilitate transfer of more complex, innovative knowledge. In innovation management, the mandated involvement of employees in temporary project teams has been much studied in the recent past, and shown to contribute to innovative performance (e.g. Cooper and Kleinschmidt 1986).

The informal network

Informal relations between individuals augment formal relations in getting things done (Lazega and Pattison 1999). Blau and Scott (1962) observed that it is impossible to understand processes within the formal organization without investigating the influence of the informal relations within that organization. The network of informal relations refers to the interpersonal relationships in the organization that affect decisions within it, but either are omitted from the formal scheme or are not consistent with that scheme (Simon 1976). Such relations thus relate to ongoing activities in the organization. Informal networks consist of the contacts actors have with others within the organization that are not formally mandated. These contacts are discretionary or extra-role in the sense of being initiated by individuals themselves – the informal network is the emergent pattern of interactions between individuals within organizations and the basis of shared norms, values, and beliefs (Smith-Doerr and Powell 2005; Gulati and Puranam 2009). Failing to maintain such a tie will not be a matter of negative evaluation by a superior (Gibney et al. 2009). Some have observed that when organizational issues in relation to knowledge processes are discussed in the literature, "organization primarily means informal organization" (Foss 2007; Foss et al. 2010). Culture, trust and communities of practice, rather than formal governance mechanisms, are then referred to.

The informal network provides insight into the general way "things are getting done" within the organization, possibly by-passing and sometimes undermining the formal structure (Lazega and Pattison 1999; Schulz 2003). When communication via the formal network takes too long, or when the relations required to get certain things done have not been formally established, the informal network ("the grapevine") may come into play as it cuts through the formal structures and function as a "communication safety net"

(Cross et al. 2002). Even though an informal network can be elusive and not transparent and can lead to clique formation where new knowledge upsetting a status quo will not be accepted, Albrecht and Ropp (1984) suggest that employees tend to transfer new ideas with colleagues in their informal network first. Additionally, Hansen (2002) argues that informal relations allow one to tap into new knowledge more easily. Informal relations provide the opportunity for information and knowledge to flow in both vertical and horizontal directions, contributing to the overall flexibility of the organization (Cross et al. 2002; Aalbers *et al.* 2014). Informally, individuals may be willing to exchange information and favors beyond what the organization has formally mandated them to do (Dolfsma et al. 2009). This extra-role behavior can sometimes be contrary to formal instructions and expectations, but has been indicated to benefit the individuals involved as well as the organization when occurring (Bouty 2000). Informal ties have been argued to be the primary basis for the creation of interpersonal trust, which is necessary for innovative knowledge transfer at the organization level to take place in practice (Szulanski et al. 2004).

Rich ties in a multiplex network

By combining different relational aspects such multiplex, relational ties may transform into *rich* ties: when individuals are connected in a number of different ways, increased and more reliable information tends to be exchanged (Sias and Cahill 1998) – see Figure 5.1. Individuals who are connected simultaneously in different networks will have different sources of information, one source possibly compensating another (Soda and Zaheer 2012). A relation of one kind may keep in check the negative side-effects of a relation of a different kind (Marsden 1981). People are also in a better position to predict and interpret how someone will behave in one context if his behavior and attitude is known from a different context, thus reducing uncertainty (Aalbers et al. 2013). Role ambiguity is significantly reduced in the case of multiplexity as people understand better what is expected of them (Hartman and Johnson 1979). In the case of rich ties between individuals in a multiplex network, each tie is also likely to be stronger, and social capital between the individuals will be larger for that reason as well (McEvily et al. 2003). Rich ties in a multiplex network come along with advantages that are necessary for the transfer of innovative knowledge, specifically if such knowledge is socially or technically complex (Hansen 1999).

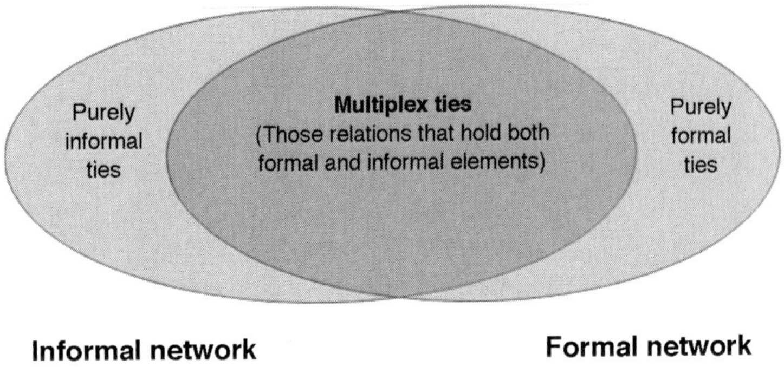

70 Networks and organization strategy

[Diagram: Venn diagram showing two overlapping ovals. Left oval labeled "Purely informal ties", right oval labeled "Purely formal ties", overlapping region labeled "Multiplex ties (Those relations that hold both formal and informal elements)". Below the left oval: "Informal network"; below the right oval: "Formal network".]

FIGURE 5.1 The informal, formal, and multiplex ties

The informal component of a rich tie in a multiplex network constitutes the trust that is necessary to be *willing* to share complex, innovative knowledge. The formal component of a rich tie in a multiplex network signifies the shared purpose and understanding and helps secure the resources necessary to be *able* to share complex, innovative knowledge on one hand. We submit that the multiplex combination of formal and informal relations in an organization's network structure in the form of rich ties proves to be a qualitatively different foundation for innovative knowledge transfer from ties that are formal- or informal-only.

Greenwood and Redrock

We base the evidence for our claims of the relevance of rich ties for innovation on our study of two separate companies, one an established European production company (Greenwood), the other a leading European financial service provider (Redrock). The two companies selected differ in terms of size, organizational design, and type of industry to indicate the robustness of our findings. Access to both companies was negotiated through the senior innovation managers, in each case operating directly under the supervision of the board of directors. Greenwood is organized according to a divisional structure (Mintzberg 1980). The company recently reorganized its activities according to a number of strategic multidisciplinary themes some years before we collected the data for this study. An organization in the financial sector, Redrock is organized as a machine bureaucracy (Mintzberg 1980). It is highly routinized, with a large operational unit and a separate unit to develop new business. Innovation activities at Redrock are focused around the theme of innovative payment methods, which is receiving significant attention by corporate management.

Table 5.1 shows the frequency of tie types in our sample in relation to knowledge transfer for Greenwood and Redrock. The majority of ties are multiplex, rather than formal- or informal-only, even though the underlying formal and informal networks measure separate networks that are theoretically independent and methodologically different, as argued above. Such frequent occurrence of ties in the formal and the formal networks between any two individuals was found by others as well (Gulati and Puranam 2009; Hansen et al. 2005; Smith-Doerr et al. 2004). Informal-only ties are, remarkably, perhaps, much less common than formal-only ties.

Combining formal and informal contacts thus, as we suggested in the theory, gives rise to advantages that do not exist when individuals only maintain a socially poorly supported uniplex tie. The appendix to this chapter provides the quantitative analysis, using methods explained at length in the methodological appendix to this book. One respondent at Redrock provides further support for this in the context of transfer of new, innovative knowledge:

FIGURE 5.2 The Innovation Network for Greenwood

TABLE 5.1 Descriptives – frequency of tie types

	Number of ties	Of which: corresponding tie in innovation network
Greenwood (114 employees*)		
Multiplex tie (‡)	116	91
Formal tie only	69	26
Informal tie only	11	6
Redrock (281 employees*)		
Multiplex tie (‡)	379	318
Formal tie only	66	34
Informal tie only	36	18

Note: ‡ Formal and informal tie overlapping between same actors; * Count of individual actors based on presence in any of the three network ties, hence deviating from number of actors depicted in Figure 2.1 (innovation only)

It was when my old mentor with whom I continued to remain acquainted informally introduced me to this group that had gathered around a new technology closely aligned to my prior experience that I got involved with the innovation community. This informal circle played out to be the basis for a formalized project that is currently developing into a new product.

Elaborating on the social antecedents for innovative knowledge transfer an employee at Redrock stated:

When I first joined the company it was hard to get drawn into the innovative activity that was going on, even though my formal role and ascribed contacts implied I was in the midst of things. It was not until I had established a place in the informal circuit that I could really connect with others and I truly got involved with innovation.

Here, the privileged access to knowledge that is more likely to be reliable is stressed. A project manager at Greenwood highlighted related benefits of trust and reduced role ambiguity from a joint informal and formal dimension as follows:

Being informally in the loop of things has helped me to be a good judge of whom to collaborate professionally with, and whom to avoid. As innovation activities are typically far from crystal clear as they unfold over time, it is important for me to know whom to bank as I develop activities into a formal project.

Another employee at Greenwood sees multiplexity in his relations as a prerequisite for involvement in the exchange of innovative knowledge, highlighting the importance of being able to predict how someone else will behave:

> When I find myself collaborating fruitfully on something really novel it generally seems to be because we understand each other and trust each other beyond just being colleagues or anything like that. … At the same time I also experience that a certain understanding of each other's professional field of work is required to really get going.

We show, quantitatively, that rich ties in a multiplex network are conducive to the transfer of new, innovative knowledge. Our qualitative data, additionally, indicates what particular benefits can be expected to be secured from the combination of both formal and informal elements into a rich tie beyond each contact in isolation.

Rich ties and innovation

Knowledge transfer is necessary to increase the innovative potential of an organization, contributing to its dynamic capabilities in a turbulent economy. Informal relations in particular have been emphasized as contributing to knowledge transfer, while formal connections received less attention in the literature since the late 1980s (Cross et al. 2002; Damanpour 1991; Stevenson and Gilly 1991). We compared how much these different networks separately contribute to innovative knowledge transfer within an organization (Hansen and Lovas 2004; Hansen et al. 2005), but importantly also establish how much they contribute when combined in a single relation between any two individuals.

Rich are not just so because multiple dimensions of a relation are combined, but they may be referred to as *rich* particularly because they contribute significantly to the transfer of new, innovative knowledge. Rich ties in a multiplex network drive the transfer of new, innovative knowledge transfer more than formal-only or informal-only networks. These insights add an important element to the research of what drives the transfer of new, innovative knowledge (Ibarra 1993; Burt et al. 1998; Tsai 2001; Kalish and Robins, 2006; Teigland and Wasko, 2009; Whelan et al. 2011).

- It is not just informal relations that contribute to innovative knowledge transfer: formal relations are a substantial driver of transfer of new, innovative knowledge as well.
- Relations that combine formal as well as informal aspects into a single relation between two persons have a genuinely distinct and significantly positive effect on innovative knowledge transfer within organizations.
- These insights can be used as a basis for ubiquitous management interventions generating positive outcomes for the organization and its members.
- These insights provide guidance to management on how to intervene to contribute to an organization's innovative capability, one of management's prime strategic objectives (Dyer et al. 2011).

Since formal relations are typically more purposefully malleable than informal ones, and as formal relations may provide the basis on which informal relations develop (Han 1996) to form rich ties, management may actively seek to enhance an organization's innovative capabilities by purposively shaping the formal structures in their organization. Management can influence knowledge transfer more purposefully than much previous research emphasizing informal relations has led scholars and managers to believe.

Rich ties can be fostered, for instance enhancing the intended effects of a management intervention. Existing informal connections can for instance be formalized ("enriched"), banking on the effect of multiplexity. More often, however, formal connections can be established where no ties existed before, providing a basis for informal connections to subsequently develop as well. When people have already established both formal and informal connections, the formal connections can be altered in a strategic move to re-align the business, relying on the ongoing informal connections to allow the organization to have its cake and eat it. Newly shaped formal ties help people within the organization to connect to different people than before, exchanging relevant knowledge, while extant informal ties still allow them to connect with others one already knew, perhaps across unit boundaries, when needed. Moving from a discipline-orientation to a market-orientation, for instance, can then prevent developing solutions twice for different markets.

Management can only intervene successfully, however, if they have a full view of all the relevant networks in an organization, and understand how these interact. This book provides insights about what should work, and why, while the methodological appendix provides insights into how one is to generate organization-specific insights required when managers are in need of in depth understanding of how their organization works, perhaps in preparation for an intervention. Managers can thus, prominently, seek to alter existing network structures and communication patterns to facilitate the innovative activity within the organization. A board member of one of the companies we studied argued that:

> No manager can truly see everything that is going on at the shop floor. But being able to identify who are in the midst of things [offered by a network perspective of an organization] really helps in not losing touch with what will shape the future of our company.

Main take-aways for Chapter 5

- Managers can make a contribution to an organization's innovativeness as manager, shaping formal connections.
- But particularly when they understand their organization's multiple networked collaboration patterns.
- So they cannot just create, strengthen and sustain links where needed, but also make use of overlaps between networks establishing multiplex, innovation-rich ties between employees.

Appendix to Chapter 5

The methodological appendix to this book explains how data to perform this analysis is to be collected and analyzed. We present the outcome of our analysis below, referring to it in the main body of this chapter. Table 5A.1 presents the results of our QAP (Quadratic Assignment Procedure) analysis of the influence of different kinds of relations on innovative knowledge transfer for Greenwood and Redrock separately. Models I and II analyze the influence on knowledge transfer of the informal and, separately, the formal network structure. In model III we include both the formal and the informal networks as independent variables to again explain the innovative knowledge transfer network as our dependent variable. Results in Table 5A.1, in models I and II, show that both the formal and the informal relations each, separately, explain innovative knowledge transfer in an organization. What may be remarkable, looking at models I, II and III, is that betas for the formal network appear to remain larger than for the informal network.

TABLE 5A.1 Innovative knowledge transfer in organizations – QAP regressions

Type network	Model-I	Model-II	Model-III	Model-IV [‡]
Greenwood				
Informal	0.704***	–	0.369***	0.137***
Formal	–	0.722***	0.444***	0.283***
Multiplex	–	–	–	0.697***
R^2 (adj.)	0.50	0.52	0.58	0.58
Redrock				
Informal	0.803***	–	0.329***	0.215***
Formal	–	0.844***	0.572***	0.155***
Multiplex	–	–	–	0.836***
R^2 (adj.)	0.64	0.71	0.75	0.77

Note: QAP semi partial regressions (UCINET; Borgatti *et al.* 2002). Coefficients standardized; 5000 permutations; *** 1% significance. [‡] Formal-only and informal-only relations net of multiplex relations.

Model IV includes the multiplex (rich), formal-only and informal-only networks to explain the shape of the innovation network. Model IV shows that the multiplex network of rich ties is better at explaining the shape of the innovation network than ties in the other, formal-only and informal-only networks. Thus, contrary to what is generally acknowledged in both the management literature as well as seems adhered to in management practice, the multiplex network of rich ties, combining both formal and informal aspects in a relation between two individuals, is thus particularly fruitful for innovative knowledge transfer.

Note

*   The chapter draws in part on joint work with Otto Koppius previously published in the *British Journal of Management*. (Aalbers et al. 2014).

6
CROSS-TIES FOR INNOVATIVE TEAMS*

Project teams have long been an essential instrument to accomplish organizational objectives (Ancona and Caldwell 1992a; Blindenbach-Driessen et al. 2010). Most organizations set up temporary project teams as a matter of course. One study reports that over 80 percent of Fortune 1,000 companies are organized around temporary teams (Garvey 2002). Project teams are set up in a plurality of context, varying from aerospace engineering and biotech companies, to more conservative financial service organizations and hospitals. Project teams swiftly undertake non-routine tasks. Teams have access to diverse knowledge domains, assess their knowledge more accurately, are better at detecting errors and biases, more likely to consider different perspectives, coordinate more efficiently than individuals, and team members motivate and help each other (Cummings, 2004; Singh 2005; Ding et al. 2010; Singh and Fleming 2010). Collective creative insights can emerge during social interactions with team members (Hargadon and Bechky 2006). Having different perspectives eliminates unhelpful insights early on (Singh and Fleming 2010) and rather allow for accumulation of knowledge from the start (Girotra et al. 2010).

Stimulating innovation, management can either assemble teams or encourage their bottom-up emergence. An example of the former is a design and NPD (New Product Development) crew, one example of the latter is a brainstorming group that meets regularly for creative ideation. Teams allow pooling different knowledge, social networks, and skills (Cummings 2004;

Singh 2005; Ding et al. 2010; Singh and Fleming 2010). These temporary work teams bring together employees from different parts of an organization to accomplish a specific task. Employees brought together in teams tend to have a diverse background.

Managing these temporary project teams poses a number of challenges. Taking a network view of the organization will help understand and address these challenges, however. This chapter takes on the issue of challenges that arise from managing projects teams, from the perspective of team members, team leaders and general management, head on. Leveraging the organizational networks to the benefit of these project teams offers key insights.

While it will by now be clear to the reader that many insights from a social network view of the organization can be used, here we focus on a particular one: the extent to which team members maintain contacts outside of the project team. A project team may be well connected horizontally so that it may tap into the knowledge base of other departments, unlocking a diversity of input required to come up with innovative outputs. At the same time being well connected vertically to upper levels of management helps project teams in finding support, buy-in from leadership, and fits with the activities in the rest of the organization. Recent research by the authors established that especially the latter, vertical ties, benefit a project team as its members leverage hierarchical relations. While such ties should not necessarily be concentrated with one individual, they should not be spread too thinly either.

Innovative project teams and networks

Companies tend to organize their innovation endeavors in project teams that are typically of a multi-disciplinary nature (Griffin 1997). Organizations must deal with increasingly complex, technical knowledge from different backgrounds, and so need to bring together their best people to tackle the issues involved in teams where different knowledge, social networks, and skills are pooled (Cummings 2004; Singh 2005; Ding et al. 2010; Singh and Fleming 2010). For managers, teams have long been an essential instrument to accomplish organizational objectives (Ancona and Caldwell 1992a; Blindenbach-Driessen et al. 2010; Haas 2010; Kratzer et al. 2010; Leenders et al. 2007a; Markham 1998). Teams are set up to pursue innovation because innovation involves generation of truly novel and useful ideas – ideas that are more likely to emerge when members build on each other's diverse contributions (e.g., Kohn et al. 2011; Baer et al. 2010). These innovation teams require the employees involved to be on top of their game, applying their knowledge to address a multi-dimensional problem. In a team they must at the same time collaborate with other experts. Additionally, they are temporarily supervised by someone who they may not know and who may not fully appreciate the knowledge they bring to the team.

How a project team works and can be managed has received some attention in board rooms as well as in the management literature (e.g. Haas 2010; Kratzer et al. 2010; Leenders et al. 2007a; Markham 1998). However, the failure rate of innovative project teams is high (e.g., Sivadas and Dwyer 2000). Approximately only one in ten product concepts developed in teams succeeds commercially (Cooper et al. 2004). Much can be gained when innovation projects can be made more successful.

Here, we focus on "cross-ties," i.e., the external ties maintained by a team within the company, showing how such ties critically enhance project performance (Ancona 1990; Ancona and Caldwell 1992a; Marrone et al. 2007). A network view of the organization emphasizes how access to diverse knowledge and information provided by cross-ties from the team to other parts of the organization are critical for project team performance and innovativeness (Blindenbach-Driessen and Van den Ende 2010). Access to diverse knowledge and insights yields better informed decisions and helps a team benchmark their activities and enhance their functional expertise (Haas 2010; Roth and Kostova 2003; Szulanski 1996). In addition to *horizontal* cross-ties, we distinguish *vertical* cross-ties. Distinguishing between horizontal and vertical cross-ties is relevant both from the perspective of management literature as well as for managers.

How cross-hierarchy ties and cross-unit ties contribute to innovation

A new product development team's external connectedness, that is, the maintenance of connections with others beyond the relations among team members themselves, has scarcely been studied (cf. Marrone et al. 2007; Marrone 2010). Connections with others outside of the team is important for multiple reasons. Teams are not autonomous entities within organizations. Teams need to manage the degree to which their activities align with the expectations and output of others in the organization. Teams are governed, if only because what they deliver will be part of a larger whole. Teams are no islands on their own, they benefit and thrive by interaction with the rest of the organization. Isolation is likely to hamper the effectiveness of team

Crossing formally and informally

Organizations contain a potentially large number of networks in them. In the introduction we have argued that two important ones among these are the formal (workflow) and the informal networks. Cross-ties in each of these two different networks are likely to offer distinct benefits for project teams. Formal cross-ties are different from informal cross-ties, and can have very different consequences. Figure 6.1 indicates this: connections emerge for different reasons, under different circumstances and have a different rationale. Who are connected in one network can be different from who are connected in another network. What may be expected of them in terms of contribution to knowledge transfer, innovation, or contribution to team performance is different as well.

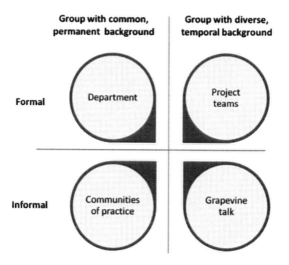

FIGURE 6.1 The formal vs. informal elements in an organization

innovation (Haas 2010; March 1991). This chapter explicitly and conceptually distinguishes between horizontal ties crossing unit-boundaries and vertical ties crossing hierarchical boundaries, within an organization. Horizontal cross-ties do provide a team benefit because the diversity of information and knowledge that is available to the team which allow it to be more innovative. Vertical cross-hierarchy ties, however, provide access to (political) influence that assists the team by finding support and resources (Atuahene-Gima and Evangelista 2000: 1269; Haas 2010). A team, acting in an uncertain and ambiguous technical, market and organization environment, to be successful (Frost and Egri 1991; Maute and Locander 1994) requires both horizontal cross-unit as well as vertical cross-hierarchy ties.

At the same time, many of today's challenges for organizations are non-routine. Project teams need to interact, structurally, yet efficiently and effectively. These interactions may be of the formal or informal kind, and may encompass ties crossing unit-boundaries and vertical ties crossing hierarchical boundaries, within an organization. The contribution to performance is different between horizontal and vertical cross-ties. Figure 6.2 indicates this.

Successful innovation teams concentrate horizontal and particularly vertical cross-hierarchy ties with a small number of team members rather than have them scattered across a large number of team members.

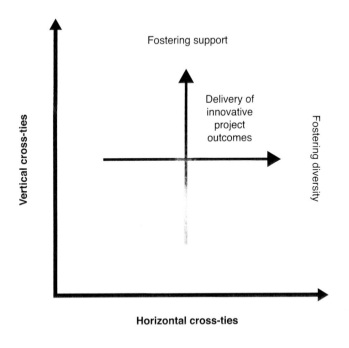

FIGURE 6.2 Horizontal and vertical cross-ties

Relevance of horizontal ties

Through effective communication, using the knowledge developed by others outside the team, teams obtain previously unavailable information and can then develop new knowledge and insights (Sethia 1995; Moenaert et al. 2000). When shared within the project team, the diversity of insights and knowledge benefits the overall project team's knowledge base and hence team performance (Allen 1977; Tushman 1979; Ancona and Caldwell 1992b). Figure 6.5 on page 91, illustrates this.

For the team to be creative and develop novel and useable solutions to technical and commercial problems, interaction and cross-fertilization of ideas beyond team boundaries can be essential (Leenders et al. 2003). Through consultation and interaction, teams may anticipate and prevent potential weaknesses in technical and marketing solutions. Communication crossing team boundaries makes it possible to access external knowledge, to be combined into new knowledge and insight. The performance of an innovation team consequently depends in part on the team's communication effectiveness. These efforts may be of the formal as well as of the informal kind. Teams that do not communicate effectively beyond team boundaries with outside specialists are less likely to generate feasible novel solutions to the multifaceted problems organizations face.

Accessing knowledge from across organizational boundaries is an important driver of innovative performance and project team success (Cohen and Levinthal 1990; Obstfeld 2005; Leenders et al. 2007b; Tortoriello and Krackhardt 2010). Besides bringing in their own specialized expertise, team members who maintain horizontal cross-unit ties to other business units are more likely to think and act outside of the narrow confines of their own task and project team (Duncan 1976; Floyd and Lane 2000). Having access to diverse resources stimulates creativity in itself (Woodman et al. 1993; Paulus 2000; Reagans and McEvily 2003). Complementary functional expertise may be brought to bear; participation in cross-unit activity by members of an innovation team increases access for the team to alternative ideas and insights (Floyd and Lane 2000).

Teams that pay attention to the crossing of horizontal ties secure access to the diverse knowledge domains of their members, assess their knowledge more accurately, are forced to consider different perspectives, and benefits from enhanced motivation as team members motivate and help each other (Cummings 2004; Singh 2005; Ding et al. 2010; Singh and Fleming 2010). At a collective level creative insights emerge during and due to the social interactions with team members (Hargadon and Bechky 2006). Having different perspectives additionally has been found to eliminate bad insights early on (Singh and Fleming 2010) and helps build on each other's ideas towards more substantial and useful innovations (Girotra et al. 2010). The following project case illustrates this.

> ### Project A: Innovation moving forwards
>
> Project team A, where 30 individuals collaborate, develops a futuristic business channel – boosting customer intimacy as well as operational efficiency. Team members are well aware of each other's tasks and responsibilities, although the exact project scoping is still less clear *Particularly now the project is becoming more visible to higher management, the sense of urgency stimulates people to follow on and share their knowledge.* With several technical specialists closely cooperating with highly connected managerial colleagues, the team seems well connected within the organization. Given some major technical challenges in the project's scope, team members have already started calling upon their personal relations to make sure all expertise available is put to use. Team morale seems to be high and so are internal team expectations of each other and of the final project result.

Relevance of vertical ties

Hershock et al. (1994) argue that continued senior management commitment and support is a most important factor in increasing the likelihood of project team success. Vertical cross-hierarchy ties connect the team to individuals with higher status positions that have desirable influence resources including access to funding, prestige, power, and privileged access to others in the organization. Although the relationship between upward influencing capability and performance is not new at the individual level of analysis (Athanassiades 1973; Porter et al. 1981; Schilit 1986), studying the capability of upward influencing at the project team level has remained largely unexplored. The limited number of studies that have researched the project level, focus on the project team leader specifically (Shim and Lee 2001) and emphasize the influence of a single manager on his subordinates (Tourish and Pinnington 2002).

Taking the team perspective as point of departure, the chapter poses that besides access to a broader range of information, cross-hierarchy ties also provide a project team with the capability of upward influencing power enhancing project team performance. Figure 6.6 on page 92, illustrates this.

Vertical cross-hierarchy ties can provide the team with access to resources of a different nature than that which the team accesses through its horizontal cross-unit ties – especially influence resources that may not be commonly accessible to the lower echelons in an organization. Teams that have cross-hierarchy ties may be expected to have access to information and other resources that provide them with a broader perspective than those who do not have vertical cross-hierarchy ties (Cross and Cummings 2004). Project A illustrates the relevance of vertical ties for successful team innovation.

Project A: Innovation moving forwards (ctd.)

The futuristic nature of the project deliverable of project A has created awareness about securing managerial involvement, a task that is trusted to two of the more tenured team members that are known to be connected well. This is what the team leader says about this:

> It is vital to know how to use my contacts and tenure to get ahead of the pack and to secure capacity for our pilots or proofs of concept. [...] My colleagues know that and respect this as it helps us to move forwards.

This seems to be successful in the eyes of team members:

> Management is clearly involved with our business. I believe our project team manager has helped in getting them there and getting us involved too. I have seen that differently at other projects", and "[t]he number of stripes does matter in our organization. We have only a few of us who can really make these stripes work to our advantage. Our project team manager is one of those people.

> **Project B: Trendsetting**
>
> Project B anticipates one of the major trends identified in the market. The project, involving 20 employees, was greeted with great enthusiasm by team members as well as by corporate management, and the team seems to have secured an effective way of raising awareness among peers and keeping people involved. Although the project team is the smallest of the teams that are classified as successful, it seems to have managed an effective division of labor. Still, the team believes more is to be made of the team potential, and relationships with key stakeholders within the company are revisited to assure fit of the team with company objectives. The team's communication network is seen by team member as one of its important assets.

For project B, cross-hierarchy ties allowed the team to gain a perspective of how the team output fits in with the organizations' overall objectives and goals. Empirical research has shown that access to higher hierarchical levels helps teams to take stock of what is relevant from a technical or commercial point of view within the rest of the project or organization so team activities can be aligned to this (Hansen et al. 2001; Nahapiet and Ghoshal 1998; Subramaniam and Youndt 2005; Mom et al. 2009). Teams without such a view may tend to focus on their isolated part of the overall design task, neglecting the bigger picture. Teams that utilize cross-hierarchy ties also gain access to support and influence resources (Blindenbach-Driessen and Van den Ende 2010). The higher hierarchical echelons in the organization provide legitimacy to information obtained to either a person or an idea, helping teams to put their plans into action (Brass 1984; Cross et al. 2001). Access to influencers can help in bringing new ideas developed by the innovation team to the positive attention of management, it can generate positive publicity, and it can even hamper or stop competing projects (Bonner et al. 2002; Kijkuit and Van den Ende 2007).

Cross-hierarchy ties can help the team resist efforts by management to impose inappropriate agendas and prevent extensive debate over aspects of and constraints for their projects (Haas 2010). As organizational politics may not be the strong suit of innovation professionals, having a champion can positively affect the team's performance (Markham 1998; Kelley and Lee 2010; Weissenberger-Eibl and Teufel 2011). Cross-hierarchy ties thus provide innovative teams with management-related resources that assist them in performing their tasks. Although some previous research has looked into categorizing boundary spanning activities (Ancona and Caldwell 1992a; 1990), no strict conceptual distinction has been made between horizontal and vertical boundary spanning activity.

Horizontal and vertical cross-ties each provide the team with distinct resources that can enhance a team's innovative performance in distinct yet complementary ways. Figure 6.3 summarizes these relations.

More may not be better: network efficiency

Making sure horizontal and vertical cross-ties are available to the project team benefits innovative performance. Yet it goes without saying that the number of ties maintained, horizontally or vertically, cannot expand indefinitely. Employees with a large number of established relations are known to strongly rely on these and are known to ignore opportunities for initiating relationships with new partners (Gulati 1995; Tsai 2000, 2001). This behavior is due to the costs involved in establishing and maintaining relationships (Tsai 2000), and may be expected to apply to the team level as well. Time spent searching for, and transferring knowledge from sources outside a person's own established network takes time away from working on one's own functional tasks (Haas and Hansen 2005). In line with Haas and Hansen (2005) one expects that incurring such search and transfer costs is worthwhile if there is substantial learning, resources or political

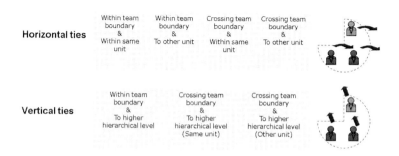

FIGURE 6.3 Horizontal and vertical ties

support to be gained, but when benefits are marginal or negligible due to redundancy in ties, actors are likely to channel their time to more economically profitable activities.

For teams to utilize both types of ties effectively, we suggest that cross-ties should *not* be scattered across the team, with most of the team members being involved in maintaining external relations. Rather, external ties are preferably maintained by a limited number of team members only. Allen (1977) was among the first to stress how specialization at the innovative team level enhances the flow of knowledge and thus stimulates scientific and technological developments. The boundary spanner, receiving only modest attention in the literature in recent years, is a key factor in the innovation process. Boundary spanners acquire, translate, and disseminate external resources throughout the organization (Whelan et al. 2011). Although earlier studies commonly leave out this distribution of boundary spanning activity among team members, one would expect that a balanced distribution is important for a project team to function effectively. Previous research has also generally assumed that vertical cross-ties are maintained by one individual only. We argue that the number of vertical cross-ties for successful innovative teams can be maintained by more than a single individual. For successful teams, they are likely to be concentrated with a limited number of individuals. In addition, vertical cross-ties have to be more concentrated than their horizontal cross-ties.

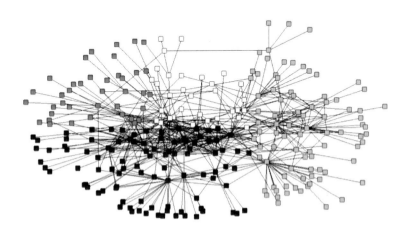

FIGURE 6.4 Innovation network at Redrock

Empirical evidence: a leading financial service provider

As a leading European financial service provider, Redrock depends on reliable technology and processes, and supports this with investments in product and service innovation. Redrock is organized according to a unit structure, following a functional segmentation, with much autonomy for the separate units. Figure 6.4 presents Redrock's vibrant innovation network. All individuals involved in innovation, either as part of a project team or who are involved in other organizational units are included. Node shape indicates business unit membership.

Horizontal cross-ties

All members of the organization were remarkably capable of identifying the project teams that were successful and related that to the teams' ability of incorporating the insights of peers that were not official team members (i.e. horizontal cross-ties). The under-performing teams were commonly perceived as much less connected horizontally. As a member of one of these teams put it:

> Everyone is aware of the benefits of scouting new ideas and getting others involved, yet ideas and talents are being wasted. We lack effective distribution of our ideas to colleagues outside of the team or Innovation unit.

This perception clearly contrasts with the observation of a senior member of a highly successful team, who notes:

> Much of our expertise lies in knowing who is doing what inside the firm. When we need it, we can get it.

A colleague member of his team adds:

> By means of my formal and informal contacts I believe to have a rather good understanding of what goes on within the organization and whom to approach to get things done for my project.

There is a clear tendency for team members of both performing and under-performing teams to try to include colleagues from outside of the team in

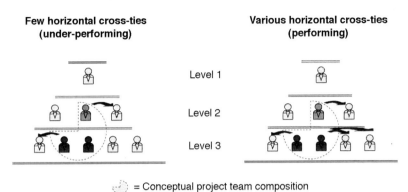

FIGURE 6.5 Horizontal ties' contribution to team performance

their innovative activities. Recurring themes brought up by the interviewees related to the diversity of insights, specialized expertise, back-up in case of unforeseen events such as illness or job transfer by team members and sustainability of the final project deliverable in the organization. All interviewed project leaders and project members raised the topic of horizontal knowledge themselves, indicating the salience of such ties to them. Members of performing project teams linked having sufficient horizontal ties to situations where they were allowed to think and act outside of the boundaries of their individual task – they indicated that this greatly benefited the performance of the team. A manager of one of these teams clarifies:

> Historically we actually have quite some contacts on our own when it comes to other fields of specialty relevant to our project. I became more aware to utilize mine to our advantage. As a result, involvement was created with other specialist within the company which has led to improvements in the conceptual design.

Successful horizontal relations seem to be of a reciprocal nature, meaning that help and insights are bidirectional, favors are returned. As one of the team members of a recently launched innovative project team signals, the "water cooler effect" is not just a conceptual notion, but actually works. According to one of the senior project team managers:

> Our expertise is appreciated throughout the organization and we can use this to our advantage when looking for input ourselves", there is

certainly sufficient sharing of ideas, for instance at the coffee corner and in unit meetings in which we partake to structurally update on our progress to our division management.

Members of unsuccessful project teams were also aware of the relevance of horizontal cross-ties, but were unable to organize these effectively. The director of the innovation unit overseeing the innovative projects remarks about the under-performing teams that these:

> are far too much internally focused, trying to get it right by themselves, and they fail to get others involved.... Clear coordination is also lacking.

The director adds that one of the under-performing projects displays a team structure that is:

> getting stuck in attempts to distribute ideas within the team. These efforts seem to be largely failing, however, and opportunities identified by some team members are not considered, let alone exploited by the project team to really get things going. This demotivates team members and leaves only a handful of individual to get the project going.

Maintaining sufficient horizontal cross-ties proves fruitful to the project teams observed. Figure 6.5 graphically summarizes these observations, contrasting successful team structures with the less victorious ones.

Vertical cross-ties

Performing innovative teams have considerably more cross-hierarchical ties than under-performing teams. The number of vertical cross-ties that skip at least two hierarchical levels is substantially higher for the better performing teams (43.0) than for under-performing teams (29.0). The number of ties directly to senior management, the highest echelon of the organization, on average, is substantially higher for performing teams than for under-performing ones (19.3 versus 13.0).

Interviews with management provide further insight: management clearly recognized that the most successful project teams were well connected to upper management and had secured a champion and other political support. Interviews with management indicate that teams with limited vertical ties were more vulnerable to being terminated in the early project stages.

92 Networks and organization strategy

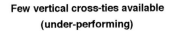

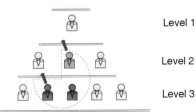

◌ = Conceptual project team composition

FIGURE 6.6 Vertical ties' contribution to team performance

Reflection by the management team members on a decade of experience with innovation projects further indicated that the innovative output of teams that had sufficient vertical cross-ties was more likely to be successfully implemented in the organization's operating core.

The following observation by a team member of one of the successful teams summarizes the overall sentiment effectively:

> Being able to utilize the established relationships with higher echelon management by a number of team members has helped [the team] to secure critical resources to prove our value to the company.

A colleague from one of the other "performing" teams added that:

> Management is clearly involved with our business. I believe [our project team Manager] has helped in getting them there and getting us involved too. I have seen that differently at other projects.

Differentiated benefits of cross-ties

Besides highlighting the effects of cross-ties in themselves, horizontal and vertical cross-ties provide teams with *distinct* benefits. Both the questionnaire and the interviews revealed that horizontal cross-ties mainly provide diversity of knowledge and ideas to the teams. Team members indicated that horizontal contacts raised their awareness of alternative insights and new ideas that were valuable to team objectives. Members of the performing teams asserted that horizontal cross-ties had stimulated them to think creatively and explore new avenues to finding (technical) solutions and had provided them with creative stimulus that was clearly different from their regular, more routinized, intra-team approach to their development task. The unsuccessful teams were aware of the relevance of horizontal cross-ties, but were unable to organize these effectively, and expressed that this had especially hurt their solution-finding success.

> ### Project C: Challenges remain
>
> Project team C is a project team in turmoil. The team is smaller than the others with ten members, but is certainly not the least communicative team. Although already "on the road" for a while, team members criticize the unclear scope and insufficient information being shared within the team: some play their relations quite close to the chest. "If they do so, I might as well do so" is how one team members responds to this. Interaction with other parties within, but also outside the organization is described as rather poor.
>
> > Everyone is aware of the benefits of scouting new ideas and getting others involved, yet ideas and talents are being wasted. We lack effective distribution of our ideas to colleagues outside of the team or Innovation unit.
>
> Team members as well as the team leader point out that corporate management does not seem to be much involved ("There is little communication between higher management and the rest of the company"), and several team members believe that the project as under-prioritized by management ("It seems as if management is not committed to us; gaining access to higher management seems not to happen"). One team member, fearing for how this experience might impact his career at the company considers:
>
> > This project might be stopped next year, if things continue as they go at the moment. I might as well bail out now as management does not seem to notice what we do too much anyway.

> But alas, management does not concur. Several attempts to increase the involvement of others have failed for a variety of reasons, which has resulted in an internal focus by the majority of the team members, despite an acknowledgment that this is inappropriate, by the team leader ("There is insufficient between-teams talk about innovation"). Although relevance of the project and its innovative contribution (generating a new product channel) are still seen as evident, morale seems to be rather low. One member of the team complains in general, reflecting the mood of others:
>
>> Nobody in this team takes charge or seems to look at the bigger picture; everybody is taking care of their own immediate interests only,
>
> and as a result:
>
>> [t]hings look poor; nothing seems to get done and nothing is accomplished for production to take up. It appears that no one in the rest of the organization is considering cooperation with us.
>
> The team leader concurs:
>
>> The project is in a pilot phase with low support within the organization and low resources to increase this support.

Vertical cross-ties, on the other hand, were reported by team members to principally provide their teams with access to managerial influence and organization-related information. The benefits of sponsorship, managerial awareness, and a sense of relevance and direction were mentioned by members of the successful teams and by overall management. Under-performing teams were not capable of securing these benefits even though they were keenly aware of the advantages of access to influence resources. Hence they displayed frustration at their own team failing to attract these particular resources.

In summary, horizontal cross-unit and vertical cross-hierarchy ties both enable innovation teams to be more successful. These ties do so by providing distinct resources: horizontal cross-ties mainly provide access to knowledge and information that is substantively tied to the team´s task, whereas vertical cross-ties provide the team with managerial support and information that promotes the team's chances of survival.

Concentrated horizontal and vertical cross-ties

For the successful teams, few team members tended to maintain at least half of the team's horizontal cross-ties. For the two under-performing teams these proportions are higher: horizontal cross-ties are less concentrated in the under-performing than in the performing teams. In the words of a team member of one of the successful teams:

> Responsibilities are clearly defined within our team. Some are better at talking to management, others are plain specialist who get us noticed in another manner – and make sure we are recognized by others (specialists). Both make us successful as a team.

He continues to add that:

> It is vital to know how to use my contacts and tenure to get ahead of the pack and to secure capacity for our pilots (proof of concepts). …My colleagues know that and respect this as it helps us to move forwards.

All teams in our sample have concentrated their vertical ties more strongly than their horizontal ties. Vertical cross-ties show a similar picture. For performing teams, half of the vertical cross-ties were maintained by less than 13 percent of the team members, while for under-performing teams vertical cross-ties were more spread. When the majority of cross-ties is maintained by only a few team members this frees the larger part of the team from having to deal with the maintenance of ties outside of the team. Maintaining such ties tends to be costly, and may prevent team members from focusing on ongoing work.

Interviewees also brought up that performing teams might concentrate both their horizontal and vertical ties among a small number of team members. Specialization regarding relationship management with the higher echelons is repeatedly related to a better functioning team. Observations show that members of the performing teams had clear views and expectations about each team member and their strengths and weaknesses, including in regards to management activities. Team members of the performing teams clearly articulated the benefits of this division of labor to enhance performance, utilizing skills of each individual effectively, and to keep team morale high. Team members who proactively developed and maintained horizontal or vertical cross-ties were perceived positively by colleagues and senior management, who referred to them as "entrepreneurs," "experts" or "organizational runner-ups". Management, in turn, appreciated only having to maintain contact with a limited number of representatives from a team, rather than being approached by a larger group. Under-performing teams, in comparison, were much less clear about role distribution, an important reason for low morale and conflict in the teams. Effective team coordination was lacking, something that was mirrored in what management observed. Trying to compensate for the lack of horizontal and vertical coordination, more individuals began to be involved in horizontal cross-ties and especially vertical cross-ties, however only increasing frustration among team members as well as management.

Project D: Failure is an option

Project D addresses a market opportunity derived from recent developments in a market related to the current market for Beta, seeking to apply core competencies in a novel way. It has confronted some major hurdles, however.

> Since we have no common goals and leadership, all seems to face much resistance.

Several of these hurdles are related, according to team members, to the way in which the team taps into corporate resources and engenders managerial commitment. "Why can we not connect to the right sponsors?" laments one team member. The team felt hard-pressed to stay on top of the game. Each team member seems to be involved in deciding on the team's direction, but insights vary strongly and so decisiveness at team level is lacking. "Since people are too much involved with all kinds of things, there is a lack of focus" and "People in this project do not have clear responsibilities. The project shows insufficient innovative potential" are typical comments one hears. Members realize this, but seem unable to escape it ("Setting our own directions seems to be counterproductive as it drains energy from the team and results in a lot of debate on who should be doing what").

Main take-aways for Chapter 6

Both horizontal and vertical cross-ties contribute to a team's innovative performance – the first by tapping into relatively unfamiliar but relevant sources of knowledge and information, the latter by fostering support and fit for a project. This chapter outlined the following:

- Innovative project teams are no "isolated islands", but thrive when properly embedded in the social network environment of their organization.
- Horizontal cross-unit ties and vertical cross-hierarchy ties improve innovation project team performance.
- Horizontal and vertical cross-ties serve different purposes.
- Horizontal cross-ties provide access to diverse, task-related information and knowledge.
- Vertical cross-ties mainly provide managerial influence and organization-related information.
- Horizontal cross-unit and vertical cross-hierarchy ties should be maintained by a small number of team members for the team to perform (possibly more than 1, but not much more).

Project teams must therfore be properly formed to increase the probability of achieving success. Management should move beyond considering individual knowledge and skills. What contacts members have, horizontally as well as vertically, outside of the team, matters as well. A social network view of ties that team members bring in helps management put together appropriate projects teams.

Note

* This chapter draws in part on joint work with Roger Leenders.

PART II
Networking Interventions
Rewiring the organization

7
INTERVENING TO STIMULATE INNOVATION

Previous chapters have provided a large number of suggestions for intervention in the networks of an organization, some of them offered explicitly, but leaving ample opportunity for others that may have already occurred to the reader.

Management often intervene by setting up a taskforce with a mandate and a budget, in the form of a small formation of dedicated employees with a specific task and related activities. In this chapter we provide background knowledge for management to consider even when the specific intervention differs from the type we report upon. By discussing the specifics of a particular intervention at length, and analyzing its effects in detail, we believe that considerations relevant for other types of interventions will also become more clear. We have personally consulted with organizations about other types of interventions, with aims different from stimulating innovation as well, and could report on these in addition but that would turn out to be repetitive, not adding value to the book.

What we find here, reflected in experiences we have had in other projects, is that previously existing relations between individuals, particularly when they are multiplex (rich) are drawn on to stage the new activity that an intervention seeks to promote. Innovation ties, where new knowledge relevant to develop new services and goods, emerge between individuals who were already connected formally as well as informally. A prime instrument for management practitioners to reach strategic objectives is an organization's social networks. Direct, focused intervention is a powerful way for management to properly configure networks in their organization. Even a relatively simple intervention such as a temporary task force intervention can quite drastically change the configuration of for instance the innovation network. Even though an intervention by management is primarily directed at the mandated formal contacts between individuals in an organization forming the formal network, it can take account of the informal (un-mandated) connections that shape the informal network too. We find that employees who have not previously been involved in innovation become engaged because of the task force intervention in particular when they shared a rich tie previously. Management can thus leverage existing social relations to nurture new ties that are conducive to innovation.

Network intervention

To successfully innovate, organizations need to generate, grow and implement a sustained flow of ideas. This requires an active involvement of employees with innovation activities. The exchange of innovative knowledge that is entailed is best analyzed from the perspective of social network theory. Research on intra-organizational networks for knowledge transfer has substantially enhanced our understanding of what kind of network characteristics are favorable to innovative activity (Aalbers et al. 2013, 2014; Ibarra 1993; Burt et al. 1998; Tsai 2001, Kalish and Robins 2006; Teigland and Wasko 2009; Whelan et al. 2011). Individuals are known to be more successful at innovation, exchanging new and innovative knowledge, especially when they have more contacts, or more diverse contacts tapping into different knowledge bases (Rodan 2010; Reagans and McEvily 2003; Burt 1992).

We know little, however, about the effects of management interventions to intentionally shape the organization's structures (cf. Birkinshaw et al. 2014; Bartunek et al. 2011), in part because we know little about how organizational networks emerge, evolve, and change (Ahuja et al. 2012; Ferriani et al. 2013). This may in part be due to the sensitive nature of interventions for participants in an organization so that no research can be done, and in part to the haphazard nature of interventions so that a systematic comparison over time is impossible even when research is actually allowed. Changing the social infrastructure in an organization, a complex system where people can face competing demands and intricate interaction patterns with others, is thus challenging, both methodologically and conceptually (Doreian and Stokman 2005). A better understanding of the micro-level processes within an organization is necessary, however, and would allow management to intervene more purposefully for instance to stimulate knowledge sharing (Foss et al. 2010).

Exchange of knowledge within an organization may not occur even when all involved are aware of the need for it (Szulanski 1996). A willingness to share innovative knowledge, especially located across different parts of the organization, should also not be taken for granted (Hansen 1999: 87; Reagans and McEvily 2003; Tortoriello and Krackhardt 2010: 169). An intervention may be needed for the transfer of innovative knowledge to start or be enhanced. Management intervention in general could be considered the core of what managers do (Diehl and Stroebe 1987; DeChurch and Marks 2006).

Exchange of new, innovative knowledge is extra to the role, however, and so cannot be stimulated or even enforced directly. What actually is transferred cannot directly be controlled since much of this knowledge is tacit. In addition, since the knowledge is new, third parties such as managers cannot readily determine its quality or usefulness, and thus face a situation of information asymmetry. Management may not be aware of the knowledge potentially available to employees in an organization. Employees in turn may not know what knowledge others in the organization have or may need. As a result of this asymmetry, the exchange of innovative knowledge can be hampered. Actively and purposively managing the innovation process enhances organization performance (Cooper et al. 1999).

Because of the asymmetrical distribution of knowledge, and the extra-role nature of innovative activity, the goal of an intervention might only be reached *indirectly* by targeting elements of an organization that managers can directly affect, the *informal* but especially the *formal* structure in an organization, which should be the focus of management and of scholars alike (Foss et al. 2010; cf. Beer and Walton 1987). The formal aspect of an organization can be defined as "the planned structure for an organization" (Simon 1976: 147) and focuses on relations as mandated by corporate management specifying someone's daily functional tasks (Kilduff and Brass 2001; Rodan 2010). The formal aspect of an organization includes quasi-structures such as committees, task forces, teams, and dotted-line relationships as well as the organizational chart (Schoonhoven and Jellinek 1990: 107; Ibarra 1993: 58; Adler and Borys 1996) – it includes a task force temporarily set up to intervene in an organization. It offers relative transparency, allocates responsibility so conflict may be prevented, and can reduce ambiguity in an organization (Adler and Borys 1996). Repeated interactions through the formal organization helps to establish shared understanding between individuals (Gabarro

1990; Matusik and Heeley 2005). However, understanding processes within the organization without investigating the influence of the informal organization may be impossible (Blau and Scott 1962; Foss et al. 2010). The informal organization comprises non-mandated contacts that allow individuals to acquire information about what is going on in their organization that is of personal relevance to them (Simon 1976; Szulanski et al. 2004; Aalbers et al. 2014). The informal organization provides insight into the general way "things are getting done" within the organization (Schulz 2003), yet can be non-transparent for outsiders and a source for resistance to change, but can also be a pathway for a new mindset to become accepted (Albrecht and Ropp 1984; Hansen 2002). The informal organization may only change over the longer term and may be (far) more difficult to govern by management (Krackhardt and Hanson 1993).

Management intervention: a taskforce

Okhuysen and Eisenhardt (2002) found that a "simple intervention" can improve knowledge sharing within organizations and contribute significantly to knowledge integration. A simple intervention in an organization is a set of purposively formulated basic instructions and accompanying facilities to engage in specific behavior, executed by a dedicated temporary taskforce. We find a small dedicated taskforce to be a common *modus operandi* in firms where innovation is not completely institutionalized – often tenaciously – but nevertheless high on the strategic agenda. Typically, such a task force intervention would have a kick-off event, and then be recurrently brought to the attention of all relevant staff by those in management who support the intervention. Resources – material and non-material – should be made available for activities believed to further the goals sought to achieve, for instance by allowing for and facilitating meetings between relevant individuals (Robertson et al. 1993). A dedicated, relatively small team is tasked to stimulate a constructive exchange of knowledge and information formally, but also informally (cf. Okhuysen and Eisenhardt 2002). In this chapter we analyze how a "simple formal intervention," by middle management, by means of the establishment of a temporary taskforce, enhances individuals' involvement in the exchange of new, innovative knowledge.

A "simple" intervention differs from a complex, formal intervention such as a reorganization. Additionally, we emphasize the temporal and small scale character of this mode of managerial intervention. The effect of an intervention to stimulate innovation takes shape through both the formal and informal networks of an organization, and should then become visible from people's involvement in an organization's innovation network in which individuals participated voluntarily or non-mandated (Allen 1977; Goodwin et al. 2008; Ibarra 1993; Madhaven and Grover 1998; Rodan 2010; Aalbers et al. 2013).

As intra-organizational transfer of knowledge is a complex, two-way, iterative process, successfully intervening to enhance it can be a challenging task. Okhuysen and Eisenhardt (2002) find that the positive effects of enhanced contacts are mostly to be expected when the intervention is not perceived as threatening (cf. Shah 2000). A threatening intervention will result in "self-focus" rather than a focus on others in the group or outside of the group (Shah 2000). Okhuysen and Eisenhardt (2002) observe that three elements are typical for a simple formal intervention: information sharing, questioning others, and managing time.

While the literature suggests that management can influence both the organization's formal and informal organization, the empirical effect of managerial intervention on an organization's employees' innovative involvement, remains unclear and can be unanticipated. The effects of a formal management intervention targeted at one group may, for instance, affect other individuals in an organization than the ones targeted. We differentiate between employees who were already involved with innovation, including New Business Development (department) (NBD), or innovation unit members, and those who were not. We refer to the latter as newcomers.

In an experimental study Okhuysen and Eisenhardt (2002) found evidence that a simple formal intervention can improve knowledge integration when it leads to *windows of opportunity* for group members to consider ways to improve their work process that go beyond the formal intervention instructions (Bovasso 1996; Okhuysen 2001; Zellmer-Bruhn 2003). Intervention can cause individuals to shift their focus of attention to others in their environment, and in general stimulates interaction (Okhuysen and Eisenhardt 2002; Zellmer-Bruhn 2003; Logan and Ganster 2007). Intervention may reduce the barriers that restrict effective knowledge integration, such as lack of familiarity among individuals, distinct thought worlds, disparities in verbal skill, status differences, and physical distance (e.g. Bovasso 1996; Dougherty 1992, Eisenhardt 1989; Szulanski 1996). This point of view is supported by social cognitive models of behavior which identify an individual's social environment as an important source of information about appropriate behaviors (Bandura 1986; Porter and Lawler 1968).

People develop *routines* to behave in specific ways under specific circumstances, becoming visible in how employees spend their time. Behaving according to routines may lead one to ignore opportunities for initiating relationships with new partners (Cook 1977: 68; Tsai 2000; Gulati 1995). An intervention can create a window of opportunity to change one's behaviors and interactions in line with what the intervention signifies as important. An intervention *legitimates* certain activities over other activities (Gittell et al. 2006), since it directs attention within an organizational setting to particular themes and goals (Ocasio 1997; Logan and Ganster 2007). This holds for those whose main task it is to exchange and develop new knowledge, such as employees in an R&D lab or a New Business Development department, but perhaps even more so for employees in other departments. The latter are not focused on innovation, and might believe that they should not interfere with what is other people's business.

Intervening by simple formal interventions has been found to be particularly constructive in settings that involve *ambiguous* and/or *novel tasks*, taking down established routines and rules that restrict knowledge integration (Dougherty 1992; Okhuysen and Eisenhardt 2002). Such circumstances typify innovative environments. Drawing from earlier findings in group process literature (Henry 1995; Okhuysen and Eisenhardt 2002) we recognize that simple formal interventions can increase knowledge integration by appealing to the resourcefulness of individuals by directed encouragement.

Those who were already involved with innovation, and especially NBD employees, may find the intervention less *threatening*, since it stimulates them to do more or be better at what to them is a non-core task. The intervention may not be perceived as an implicit criticism of their previous involvement and activities. The formal intervention leads to a broadening of their functional scope through the emphasis that is put on the innovation theme. In addition, it is key that employees can see that sharing leads to immediate gains such as less hassle, or easier tasks, reduced working hours or earlier closing (McLaughlin et al. 2008).

The communication costs employees face when transferring new, innovative knowledge, typically complex in nature, are lowered by the intervention, which gives them the opportunity to add new contacts (Levine and Prietula 2011; Haas and Hansen 2005; Tsai 2000).

Those already more involved with innovation before the intervention face higher communication costs than those who are less involved. The transfer of new, complex knowledge for innovation is more likely to cross unit boundaries, increasing the costs of communication. If the intervention seeks to stimulate something that is at the core of someone's activities, a "not invented here" attitude might lead employees not to be stimulated by an intervention and also to resist outside information. When insufficient resources are made available to facilitate a particular kind of exchange, people will most likely focus on their core activities (Malik 2002; Szulanski 1996).

Intervention provides resources and opportunity for structural improvement in the innovation network. Those who, prior to the intervention, were not or less involved in the costly exchange of complex knowledge crossing unit boundaries that is required for innovation (cf. Hansen 1999; Reagans and McEvily 2003), will experience a window of opportunity due to the intervention that appears more attractive to them. NBD employees may actually be stimulated relatively less by the intervention than others.

Proper network characteristics for innovation activities

Innovation involves cooperation and relies on social interactions (Bovasso 1996). From a number of studies looking at innovation within an organization, focusing on the networks involved, it has become clear what characteristics of networks will enhance individual innovative involvement (Björk and Magnusson 2009; Ibarra 1993; Dougherty 1992; Albrecht and Hall 1991). Numbers of contacts, as well as diversity of contacts are prime among these characteristics (Tsai 2002; Perry-Smith and Shalley 2003). A simple intervention will take effect by changing (increasing) the sheer number of contacts as well as the diversity of contacts available to an individual in the innovation network where new, innovative knowledge is transferred.

Number of contacts

By communicating with others, individuals gain access to novel perspectives and unique knowledge and can generate political support for their ideas. A "law of large numbers" applies in the context of idea generation: the larger the number of sources of ideas available to an individual, the more likely one is to encounter, combine, and further develop new ideas (Ohly et al. 2010). The sheer number of ties an individual maintains relates to the ability to generate new ideas (Björk and Magnusson 2009; Ohly et al. 2010). The number of contacts an individual holds also helps in evaluating ideas according to the standards valid in a larger social context (Binnewies et al. 2007; Ohly et al. 2010). Related to this, the absolute number of relations an individual maintains correlates with the proportion of high-quality innovative ideas generated by an individual (Björk and Magnusson 2009). A large number of contacts enhances creativity and innovation because well-connected actors tend to trust each other more, are more willing to share their knowledge and ideas openly, and are well equipped to validate input received (Perry-Smith and Shalley 2003). More internal communication will thus enhance an organization's innovative performance (Foss et al. 2011). Although there are costs associated with maintaining network relationships, and hence the number of contacts cannot increase indefinitely, such a situation is unlikely to occur soon in most organizations.

However, the sheer number of ties in the innovation network maintained by individuals is not the only aspect of their contribution to the innovation network or arena.

Diverse contact, crossing unit boundaries

Diverse contacts with others provides access to diverse experiences, unique and varied resources, and alternative thought worlds (Cross and Cummings 2004; Mors 2010; Reagans and McEvily 2003). Holding cross-unit contacts increases one's access to alternative views on an organization's existing strategy, goals, interests, time horizon, core values and emotional tone, and complementary functional expertise (Ancona and Caldwell 1992b; Cummings 2004, Mors 2010; Burt 1992; Floyd and Lane 2000). Although there are also obvious difficulties associated with transferring, integrating, and leveraging the heterogeneous inputs and diverging perspectives available across intra-organizational boundaries (Argote 1999; Carlile 2004; Dougherty 1992; Tortoriello and Krackhardt 2010), the diversity of insights can benefit the innovative knowledge base and performance of individuals and sharpen the quality and robustness of new ideas (Mors 2010; Carlile and Rebentisch 2003; Hansen 1999; Tsai 2001). Especially in low density or "non-redundant intra-organizational networks" will one be able to leverage the benefits of combining different ideas in order to create new ones (Burt 1992). In a low density or sparse network, actors are likely to receive a greater diversity of information as they relate to diverse others (Mizruchi et al. 2001; Perry-Smith and Shalley 2003). Besides bringing in their own specialized expertise and representing the interest of their own specific unit, individuals who hold diverse, cross-unit contacts also have to think and act outside the perhaps more narrow confines of what their own job and position require (Aalbers et al. 2014; Duncan 1976; Floyd and Lane 2000; Foss et al. 2011). Exposure to conflict and discussion as a result of different needs, objectives and interests between differentiated organizational units and hierarchical levels, which are likely to emerge from being involved with diverse cross-unit ties, is also believed to increase innovative outcomes at the individual level and sharpen the quality and robustness of ideas (Duncan 1976; Mom et al. 2009). When shared within the unit an individual operates, the diversity of insights and knowledge can enhance the unit's knowledge base and increase performance (Aalbers et al. 2014).

Stimulating newcomers to join the innovation arena

Prior research has indicated that intervention is an interruption of common procedure, creating a window of opportunity for people to reconsider and possibly change or add to their normal activities (Tyre and Orlikowski 1994; Okhuysen and Eisenhardt 2002). Although targeted at the established innovation community at t=1, an intervention may increase involvement with innovation by all employees, and so when only observing the effects on individuals who had been involved with innovation prior to the intervention the effects from the intervention may only be partially visible.

Individuals are more likely to become aware of other activities, and in particular those activities endorsed by the intervention, as deserving enhanced attention and being more legitimate than previously (Tyre and Orlikowski 1994; Ahuja and Katila 2004). This may stimulate the re-evaluation of the way things have been going not just by the individuals targeted. Employees who did not belong to the initial innovation community at time of the intervention will be drawn in as well.

The effects of a simple formal intervention to stimulate involvement with innovation may thus create ripple effects throughout the organization as members of the target community interact with others in the organization. They may learn independently from their contact with individuals targeted by the intervention of the potential benefits to be gained since information may take different routes to spread in a network (Newman 2003). Boundaries to being involved with the innovation community, perceived or real, have lowered due to the intervention. The legitimacy of innovation activities undertaken in the innovation network has increased, and the activities by early hour innovators reaching out will be reciprocated by newcomers (Bouty 2000; Dolfsma et al. 2009). Newcomers may be (even) less likely to perceive the intervention as an implicit criticism of their activities prior to the intervention, since it is not a core activity for them. Not responding to the intervention will not hurt them as the intervention was not explicitly targeted at them, while responding to it will give them extra kudos from management. Moreover, opportunistic behavior by newcomers may be involved as they seek to benefit from being involved in an expanding community, participating in an activity that is valued in the organization and by management (Bovasso 1996; Burt 1992). In addition to possible opportunistic behavior triggered by the potential availability of additional resources, a newcomer's intrinsic motivation to position oneself in a social setting where extra facilities are available and where participants receive prestige and recognition also contributes to the ripple effects of a simple formal intervention. Participation in the innovation arena reduces an individual's cognitive social strains and provides kudos too (Levine et al. 2001). As newcomers are not faced with high communication costs before becoming involved in the innovation network; they are likely to be less involved in the exchange of new and complex knowledge, and knowledge transfer crossing unit boundaries.

To conclude, also employees not previously involved with innovation (newcomers) may thus become aware of the purpose of the intervention and might realize the individual and organization gains that are to be had from entering the innovation arena. They can thus also be expected to alter their behavior to join the innovation community where relatively complex, innovative knowledge is exchanged even if the intervention is not targeted at them (Bovasso 1996: 1419).

Empirical evidence from Redrock: the "ripple effect"

The empirical evidence behind the claims in this chapter is based on our continued research carried out at Redrock, one of Europe's largest and most innovative financial service providers. We observed the innovation community prior to, during and after the intervention of the establishment of a taskforce, gathering evidence on the effect of this formal intervention on the social structure of the organization. The intervention involved the deployment of a dedicated taskforce to enhance innovation by increasing the relations in the innovation community through awareness creation (cf. Okhuysen and Eisenhardt 2002). The focus of the intervention was to enhance the innovation involvement of the established, early hour innovation community. While seeking to increase cooperation in the innovation network, the intervention operated in an indirect manner by arranging formal and informal meetings, in effect using the formal and informal aspects of the organization.

A Simple intervention

The taskforce to enhance innovative involvement of individual employees operated for a period of two months and was staffed by a senior and a mid-level employee, both of whom could allocate the majority of their time to implementing the intervention. Each task force member was well-connected in the organization. The intervention targeted all employees constituting the innovation community at time t=1. The target population was jointly determined by management and taskforce members. The intervention was introduced in a general kick-off meeting, and was repeatedly brought to employees' attention in bilateral and team meetings to emphasize the relevance of enhanced cooperative behavior, for both the innovation community as a whole, as well as the individual. The taskforce contacted the people identified to explain the purpose of the intervention and the activities that could be undertaken. The taskforce, for instance, offered to introduce individuals to others within the organization and facilitate the exchange that could result.

Figure 7.1 provides a visual representation of the changes in the innovation network at Redrock between t=1 and t=2. The innovation community at Redrock certainly increased in size (see Table 7.1), becoming increasingly active due to the intervention. One could also claim, as an NBD employee did at t=1, that there was a dire need for improvement: *"involvement with innovation is poor, we are truly wasting potential. Communication between NBD and the rest of the organization is at a low."*

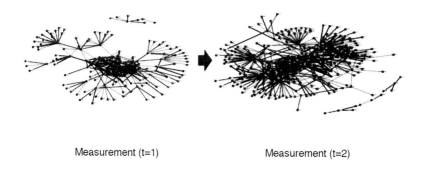

FIGURE 7.1 Innovation network, before and after an intervention (t=1 and t=2)

Table 7.1 shows a substantial increase in the number of ties due to the intervention. The increase is not due to those employees for whom innovation was the main task all along, becoming more involved with innovation. Having higher communication costs to start with does not prevent those already engaged with innovation at t=1 from adding ties upon an intervention. As one of these early innovators stated:

> I appreciate the increased buzz surrounding the topic of innovation. And honestly it is the enthusiasm of my colleagues that made me realize that there are things to be gained here, even though I am not directly responsible for innovative output, that is.

As a result of the intervention the number of cross-unit ties increased substantially as well. A manager, also already involved with innovation prior to the intervention, claimed:

> Communication regarding new ideas and services is exceedingly slow and centered around elite groups. What should be improved is discussed and developed by an in-crowd. Choices concerning innovation are made without involving relevant outsiders.

The increase for both the number of ties and the number of cross-unit ties following the intervention is high for newcomers in particular, those not engaged with innovation prior to the intervention. At t=2, not too long after the intervention that was not targeted at them specifically, newcomers had developed a substantial involvement with innovation. Newcomers to the

TABLE 7.1 Ties in the innovation network, t=1 and t=2

Employee typology	Innovation ties[‡]		Diverse (cross-unit) innovation ties[‡]	
Time period:	t=1	t=2	t=1	t=2
Early hour innovators	382 (2.65)	528 (3.66)	260 (1.87)	313 (2.25)
Newcomers	n.a.	149 (1.46)	n.a.	61 (0.6)
Total	382	677	260	319

Note: [‡] Number (average per employee).

innovation network show an average involvement that is not yet comparable to others before the intervention, but nevertheless shows a substantial jump in a short period of time from zero-involvement. A newcomer illustrates this clearly when he stated:

> I got involved with the innovation community as the result of an interesting conversation with my old mentor who introduced me to this group that had gathered around a new technology that is closely aligned to my prior experience. I actually was unaware of them running these activities on-the-side.

More people throughout the organization experienced the increased importance attached to or urgency of transfer of new, innovative knowledge. In a number of interviews this became apparent. Even in formal meetings not primarily focused on the topic, innovative collaboration is scheduled as an item for discussion, signaling commitment and importance given to the theme by management. One newcomer expressed it clearly:

> If the key players within our organization, and then I do not only mean management but also those of my colleagues that have been around here for a while, think our involvement with innovation is important, it must be.

Employees, even when they have been hired recently, sense the change in atmosphere, and also a change in the extent to which cross-unit exchange is stimulated and actually occurs, most strikingly.

The potential to grow the innovation community beyond the employees at the NBD unit and their immediate contacts is substantial. NBD usually is a relatively small unit, compared with the full size of an organization, but must, and actually can rely on other units for implementation as well as the development of new ideas. Newcomers in particular are able to become actively involved even as a result of a relatively small intervention, which is indicative of the innovation potential that can be unlocked relatively easily inside an organization outside of a unit fully dedicated to innovation activities.

Main take-aways for Chapter 7

- A straightforward, simple application of managerial discretion, in the form of a simple intervention, can significantly improve the size (number of people involved, and number of contacts between them) and quality (diversity of connections, crossing unit boundaries) of the innovation network of the organization. The intervention described here at length is indicative of other types of interventions, pursuing other goals that management might consider.

- Intervention stimulates the development of the network characteristics that facilitate involvement in innovative knowledge transfer. Sheer access to others, as well as, separately, access to a diversity of others in the innovation network substantially increase due to an intervention and, surprisingly, surges among newcomers to the innovation community, employees that were *not* primarily focused on innovation beforehand and who were *not* targeted by the intervention. An intervention can thus have effects that go well beyond changes in behavior in the target population. Management should be aware of this.
- Intervention by management into the organization's network can certainly have the intended outcomes, but will have unintended ones as well. These can be positive, but, depending on the goal of the intervention, can be negative for the organization as a whole as well. One would expect that the type of intervention discussed in the next chapter, a reorganization, would have unintended consequences that are unfavorable for the organization or its management. Yet, even here, findings we can report on offer a different view.
- A simple intervention increases the number of ties in the innovation network for all, but the growth is especially noteworthy for newcomers to the innovation community.
- Network analysis opens up the perspective to a potentially untapped reservoir of un- or under-used potential within an organization.

8
INNOVATION DESPITE REORGANIZING*

Rewiring the network

Corporate restructuring by means of downsizing is a prime example of a more drastic intervention on an organizational network. Although downsizing was once viewed as an indicator of organizational decline, it has now clearly established itself as a prime mechanism of corporate restructuring (Fisher and White 2000). A radical form of corporate change, downsizing has been a managerial practice for increasing organizational efficiency and effectiveness during the last two decades (Littler 2000; Aalbers et al. 2014). The goal tends to be to improve the efficiency of an organization by decreasing costs, enhancing revenues, or increasing competitiveness (Datta et al. 2010). Downsizing may thus coincide with corporate reorganization or the planned replacement of the current organizational structure and operating model with a new, for instance more customer-centric one (Gulati and Puranam 2009). An organization's existing orientation will change as its strategy shifts relatively abruptly (Agarwal and Helfat 2009).

While at times believed to be unavoidable, corporate reorganization by downsizing is widely believed to affect innovation negatively. Downsizing dissolves social relations forcefully (Nixon et al. 2004; Fisher and White 2000), and so retaining the social infrastructure for innovation is by no means evident. Innovative capacity at the same time is a premier avenue towards corporate recovery following organizational decline (Ocasio 1995) – either of the externally imposed or self-inflicted kind. This chapter addresses how and which social networks make sure that an organization will be able to continue to meet its strategic goals. Even though a reorganization is a radical intervention, it is an instrument that managers use repeatedly. Despite its frequent use, managers have little understanding of a reorganization's actual effects on the organization's social networks. We will discuss how a reorganization affects both formal and informal networks within the organization. After a reorganization it is unknown which individuals are most likely to remain contributing to the organization's strategic goals. Being well-connected throughout the whole organization, formally and informally, will make sure people remain involved with innovation. This chapter provides managers with direction to set the scope of their individual network horizon prior to downsizing to enhance their chances of surviving the downsizing in a way that is not detrimental to their innovative capacity. It compares the tradeoff between focusing on one's direct social environment or to focus on one's position in the overall network instead. The chapter closes with a reflection of the persistence of rich connections over time and how people can stay central to the process.

Innovation despite reorganization

"Downsizing always leaves scars" (Aalbers et al. 2014: 18). Downsizing at a large European commodities trading company shocked employees. They saw colleagues called to the HR department and subsequently being escorted out of the building, without the opportunity to say goodbye, quickly collecting personal belongings. Continuing or reigniting innovative activities is not straightforward in such circumstances.

Managers need to take an end-to-end view of their innovation efforts, spotting organization-specific strengths and weaknesses, and tailoring innovation efforts in a way that is appropriate to their organization. Tough choices need to be made in tough times: to cut cost or re-align organization structure with organization strategy. In such times, innovation efforts are easiest to cut since their returns are uncertain and will only arrive in the future. Innovation is known to suffer. At the same time, however, cutting down on innovation may simply postpone the inevitable by poorly equipping an organization for future survival let alone competitive positioning. In this contribution, we point to which innovation efforts should be cut, which should be maintained, and how such innovation efforts can be maintained at a lower cost.

Finding the innovation sweet spot

Selecting among innovation efforts is important in good times, but becomes a life-saving exercise in times of crisis. In these times, investing in the development of new knowledge that may only be relevant in a distant future is no longer an option. What managers need to realize is that innovation thrives on employees closely cooperating in fine-grained social interactions. With this consideration in mind, an organization can make better choices to continue to nourish innovation despite downsizing. To determine the extent to which an organization can expect its innovativeness to recover after downsizing, the potential changes in the way in which employees collaborate need to be studied. Through collaboration, in social networks, relevant information is generated, screened and dispersed, and knowledge is developed, laying the foundation for organizational innovative capacity. (Granovetter 1973; Burt 2004).

Obviously, when knowledge being developed is both irrelevant and not supported by a team of people collaborating, an organization should abandon it as it leads to a certain death even in the best of times. When knowledge is relevant for an organization but there is no team to develop it further, the organization finds itself at a dead end. It should consider hiring experts or acquiring another organization to push forward, even in the worst of times. An organization cannot expect to develop the relevant, valuable knowledge when it has no team of experts. A team of experts may work closely together to develop innovative knowledge that nevertheless is irrelevant in the sense of not being in line with the strategy that management maintains and the (future) market demands. Such innovation activity is a dead weight for the organization. Management may, however, be lobbied by the people involved to continue developing this knowledge. The innovation sweet spot, where relevant innovative knowledge is developed by a group of experts collaborating closely, is what a manager should nurture. The question is, of course, how to nourish this innovation sweet spot, even when an organization is cutting down on resources to fund it?

The following notions form a prime point of departure when innovation-proofing a downsizing event:

- Nourish the innovation sweet spot.
- Communicate and act pro-actively about goals and procedures, to all but especially to those who constitute the organization's innovation sweet spot.

- A reactive stance lowers morale and decreases perceived fairness among the remaining employees and may make some of them leave themselves.
- A management mindful of and careful about the social networks in their organization can counter negative employee reactions in addition to preventing loss of innovative potential to the organization.

Nourish the innovation sweet spot when downsizing

When management reorganizes and downsizes innovation often takes a back seat. However, this need not be the case. Reorganization and downsizing can demotivate employees (as well as managers) and also severs links between people. Innovation, being largely something people do "extra" and also being difficult to monitor, will suffer in particular. Dougherty and Bowman (1995) suggest that downsizing hurts product innovation in particular. However, recent findings indicate that reorganization and downsizing need not bring innovation to a complete stop. Organizations in diverse industries reinvented themselves drastically, by nourishing innovation even in times of crisis, and came out on top of the game. Examples such as IBM, Boeing, Philips, and Hewlett-Packard emphasize this point. Retaining the social infrastructure for innovation is an important explanation of these successes, but at the same time difficult to achieve. An organization's innovation DNA can be the first to unravel during a reorganization. An organization's innovation DNA, the shape and dynamics of formal as well as informal relations, determine if and how an organization may expect to continue its innovative prowess. When downsizing, management should make sure that a group of individuals that had been innovating together continues to collaborate fruitfully. The network of collaborating people, an organization's innovation DNA, should be nourished. The question is, of course, how to do that.

Internally oriented brokers, but who?

Downsizing, a particularly radical form of management intervention, will have a primary effect on people's presence and activity in the formal network as functions disappear and are redesigned. The location of where people are based, and the way in which their function is defined to an extent determines with whom they interact informally as a matter of course. Thus, management can and must understand the connections present in an organization before it starts reorganizing. Actual innovation contacts can be more difficult to find. The formal contacts are most visible, and the informal ones can be relatively easily uncovered. The latter two also help management find and nourish the innovation sweet spot.

Connections between people and the overall network configuration should be what a manager who has the organization's long term success keeps in mind. Awareness of an organization's social innovation DNA impacts on recovery and success after downsizing.

A new business development manager at Redrock, for instance, faced challenges inflicted by an across the board job-cut of 30 percent. David, who joined Redrock as a technical engineer and had spent years developing a strong network throughout the company, is eager to turn innovation projects into value-adding enterprises for his company. Moving up the corporate ladder, now overseeing all new business initiatives that emerge within the organization, David realized the possible danger of the downsizing effort his company could not avoid. This would cause a significant reduction in the executive cloud for his own team, as well as disrupting the innovation relations both from his team, and to others in the company.

David noticed that some employees clearly and permanently shut down innovation activities they were involved in before the downsizing. Others continued, and even increased their innovative activities soon after the reorganization-dust had settled. And they were doing so without much managerial support. This divide within what previously had been a tight innovation department puzzled David.

Identifying innovation guardians

What we found as researchers, based on further talks with him and others and also based on a close analysis of the overall networks in the organization, was that it matters crucially how people were connected before downsizing. Those who continue their innovation activities despite the downsizing, the guardians of innovation constituting the innovation DNA in the organization, were connected more *as well as* differently in the organization's social networks. Individuals who are well-connected as well as strategically positioned individuals will continue to contribute to innovation.

The fact that the structure of connections determines continued innovation involvement means that managers can ensure innovation does not suffer from downsizing, at low cost. These guardians are *not* differently motivated or disposed towards the organization – offering bonuses for individuals who continue their innovation efforts may not have the expected result. The withdrawal or even sabotage that some show during downsizing seems mostly to be concentrated among poorly connected individuals.

Number of connections

The contribution of employees that maintained many connections in the formal (workflow) and in the informal (grapevine) networks of information flows prior to downsizing are more likely to continue innovating during and after reorganization. Particularly when any two individuals connect both formally and informally they are likely to continue their innovation efforts. They know each other well, have learned to trust one another personally, and can also make the necessary formal arrangements. Less well-connected colleagues typically gear down on the innovative activity after downsizing, in many cases reducing these to zero.

Diversity

Continued involvement with innovation can be expected from people who are well placed in the overall flow of information in an organization. Tapping into the various corners of the organization beyond their immediate contacts, these individuals receive and pass on a diversity of information, boosting their potential to make new combinations, boosting their own possibilities for novel idea creation, and improving their understanding of how their efforts fit in a larger picture. Such "betweenness benefits" are more important than those that come from immediate contacts. Better betweenness can easily compensate for fewer immediate contacts.

Having a reputation for being well-connected throughout the organization is self-sustaining: others will approach these individuals for support and to collaborate. These individuals will, as a consequence, also be more visible and are thus less likely to be the victim of downsizing. Well-connected individuals may not be visible because of their hierarchical position, and can thus be overlooked. Recent research we conducted at a downsizing organization indicates that the degree to which employees are positioned on the shortest path within the formal network prior to a downsizing event actually predicts the capacity of employees to remain involved with innovation in the event of drastic organization downsizing. Based on a longitudinal study of the social network of an organization that underwent downsizing, we examined the resilience of the innovation network following corporate downsizing. The findings are of interest to those reorganizing and the subject of a reorganization alike. As individual networks are streamlined under corporate and social uncertainty, employees that exercise control over their extended formal social environment (betweenness centrality) were found to be most capable to hold on to their innovation ties (Shah 2000). The explanation is as follows. Due to the uncertainty that comes with corporate downsizing, employees and entire organizational units may seek to strategically diffuse knowledge, perhaps in retribution to management. Employees may become reluctant to make suggestions to colleagues and information sharing can slow to a crawl (Bommer and Jalajas 1999; Gandolfi and Oster 2010). In such more conservative knowledge sharing circumstances, those that are centrally positioned prior to downsizing, in the full network (cf. Provan et al. 2007), have advantages that may be referred to as *betweenness benefits*. Such individuals receive information in larger quantities and of a larger diversity, tapping into the various corners of the organization beyond their immediate contacts. Employees fulfilling a strong betweenness position are able to interrupt or steer the flow of information that spans the whole organization.

Leaders

Next to being well positioned on the shortest path in the formal and informal network of an organization, our research shows that an individual's position on the corporate ladder ensures that someone continues to be involved in innovation following downsizing. The need for this may be better understood by leaders and middle management, as they are charged to take a broader perspective. The goals that they must meet, however implicitly formulated, align more closely with innovation as well.

Informal contacts

As formal positions might require more time to re-establish and settle, informal positions are likely to allow one to pick up on new things that could help ongoing innovative activity. Intel, for instance, when it reorganized, has purposefully relied on previously established informal contacts to support knowledge transfer in the newly shaped formal structures, at least in a transition phase.

Since informal contacts are difficult to manage, however, relying solely on these could backfire. When support for the re-aligned strategy in parts of the informal network is lacking, its weight can be used to undermine innovation efforts.

When discussing the exact way in which to downsize his NBD department with his CEO, David had specifically argued which individuals should be spared the effects of downsizing because of their position in the different social networks. Recognizing their pivotal innovation role for the company, led it to quickly reboot innovation efforts once the dust had settled.

Rewiring sensibly

Observations made earlier can be used to offer suggestions to managers to maintain innovation efforts despite downsizing. We offer suggestions focusing on issues in line with the organization network analysis we and some others have done. First of all, choices should be made about which activities should be maintained and which are preferably to be discarded. Various suggestions can be made. Irrelevant R&D efforts should be cut, but when they muster support from within the company, expect strong resistance. Developing knowledge relevant for an organization's future might require investments in hiring new employees and embedding them properly – something that might create resentment among the existing of employees who might experience uncertainty.

Existing internally oriented brokers should be nurtured in particular, so that the innovation DNA can be maintained. We suggest that key individuals may be stimulated, especially when an organization downsizes, to grow into a role of internally oriented broker. As the company downsizes and possibly also re-aligns, some redundancy in communication should be allowed for. If management is able to target internally oriented brokers in particular when it communicates its intentions, the efficiency of the downsizing and re-aligning efforts should increase substantially.

Internally oriented brokers, or potentially, internally oriented brokers, are not necessarily people higher up the hierarchy. They can and should also be individuals who are more junior. More senior employees are more likely to experience communication overload and might not provide the context of information they pass on, might be more inclined to reduce the complexity of information passed on, and may thus introduce biases and distortions. The more specific and explicit the relevant knowledge exchanged in an organization is, the less management must rely on social connections in an organization. In knowledge intensive organizations and for organizations that offer highly customized solutions to customers, in service organizations for instance, management must be very mindful that the social networks in an organization are properly formed and functioning.

We know from previous research that management can take a leading role by shaping formal structures in an organization, and that these formal structures are very likely to be the basis for subsequent informal and innovation contacts to develop. Formal contacts precede informal ones, or at least are not likely to impede their development. Managers should thus not despair: their prerogative is to shape formal structures and connections and such activity actually can be beneficial for innovation activities by employees in their organization. Managers particularly can think strategically about the contacts they would like to see grow and develop by means of the composition of temporary teams. In studies focusing on less turbulent organizational settings, managerial help in guiding employees to optimize their networks has been found to be of use for various reasons. Individual employees may find it difficult to search for relevant knowledge on their own, for instance, or have trouble integrating their knowledge with what knowledge others in the group possess. Transferring complex knowledge, especially across organization unit boundaries can be particularly challenging when no established connections exist to begin with. Independently acting on knowledge acquired, or reusing existing knowledge partly held by others to independently develop a new value proposition to the company may not be forthcoming without managerial direction to facilitate or stimulate it. Management may be able to bring focus to the networks of individual employees, as employees that boost a large number of established social relations within the organization are known to strongly rely on these and are known to ignore opportunities for initiating relationships with new partners.

When forming temporary teams there might be an inclination to bring individuals together who have worked together before, or who are the biggest expert in the field of knowledge required. An important consideration to make when forming a work team, is the inclusion of a team member who management believes should be given the opportunity to establish new connections with members that they would not have otherwise been likely to have. In particular when such newly forming relations cross intra-organization unit boundaries, horizontally as well as vertically, purposive action by managers forming a team may be the only way for connections to develop.

The fortunate thing for managers in an organization that downsizes is that we have found that motivation does not play the role in a knowledge sharing network that many would think: being intrinsically (or extrinsically) motivated will not make someone more active in the innovation network. Other management instruments such as a coherent policy and persistent commitment shown to a policy will make more of a dent. However, managerial direction aimed at establishing connections between employees that did not exist before is costly. Starting and maintaining connections will cost the company, for instance entails that time is not spent on other activities.

Under downsizing conditions, such proactive managerial involvement might not change the common notion of organizational downsizing as a strong stressor to those involved, but at the very least can turn a downsizing event into something that is not left to chance.

Main take-aways for Chapter 8

- Orchestration of downsizing is a delicate and also crucial task for management.
- Employee reactions to redundancies have been found to vary considerably depending on how the employees felt about management conduct.
- There is a natural tendency, when deciding who to retain and who to let go when an organization downsizes, to mostly look at the knowledge and capabilities that an individual holds.
- A crucial ingredient of downsizing is for management to use its understanding of the existing formal and informal networks so it can rewire connections between employees.
- Management should realize that proper understanding and use of the social networks in their organization can prevent loss of innovative potential and even counter negative employee reactions to downsizing.
- Without attention to properly retaining some wires and rewiring other social contacts, and particularly the ones that nourish innovation, downsizing becomes self-defeating.

Note

★ The chapter draws on our research previously published in the *Journal of Business Strategy*, which appeared as: Aalbers, H.L. and Dolfsma W. (2014): 18–25.

9
METHODOLOGICAL CONSIDERATIONS FOR NETWORK ANALYSIS

Networks are much talked about, in academia, in organizations, in society. Mostly, this talk is rather generic: focusing on the importance of having connections to others. Networks in this sense are referred to in rather metaphorical terms. While this is informative in itself, it is not the way in which we address networks in this book. The insights we present and discuss, and the advice we give, are all based on a thorough analysis and understanding of high quality social network data. Network data for genuine organization network analysis is more difficult to collect than other types of data that tends to be used in the social sciences. We will explain some of the challenges below. Once this data is collected, one can apply it to a number of different purposes. One of these is to produce visual representations of networks. Another is to produce descriptive summary statistics of the network, or nodes in it. For many, this is as much as is needed, but even for this purpose the data must be of high quality. Rather than discuss at length the details of network analysis and visualization, we will present some of the key concepts, ideas as well as more practical considerations for social network data collection and analysis. New techniques and new software is developed at a rapid pace – a discussion of the current state of the art is likely to be outdated in a relatively short period of time. The points we raise here, however, are central to any and all techniques for social network data collection, visualization and analysis efforts. They will help relatively novice readers, or readers who want to have a quick yet thorough overview of what organizational network analysis entails, understand what they are presented with, and even ask to-the-point questions, and determine whether an ONA (Organization Network Analysis) is performed reasonably well and if an ONA makes sense for their own organization. (Which it will.)

The data itself, once collected, also poses specific challenges when performing more advanced analysis. Since not all readers are interested in this, however, we will be brief about this and refer to the work of Wasserman and Faust (1994). The only two things about advanced statistical analysis involving network data we observe here are the following. First of all, observations in a database containing network data are not independent of each other – there is auto-correlation in the data. This means that a standard assumption in many statistical analyses, such as Ordinary Least Squares (OLS), is violated. Different analyses are needed, often involving simulation or random sampling techniques. Second, because of this, adding non-network data to network data in a single analysis can be more challenging than one might think – for some types of analysis it might actually be impossible.

Analyzing networks in organizations

Organizational network analysis demands a systematic approach to map and study the connections and resource flows between people, teams, departments and even whole organizations. In this chapter we address this systematic aspect and provide a step by step approach to conduct the analysis of an organization network. We consider basic protocol for those intending to map the network of an organization and highlight the specific requirements for analysis of an organization's innovation network, over other types of organizational networks. The second part of this chapter highlights effective network intervention methodology, and the third and final part closes with a reflection on network ethics. At a glance, the organization network analysis progresses along the following four steps:

1 Need identification
2 Data collection
3 Analysis and visualization
4 Interpretation

Step 1: Need identification

Goals and objectives

Network analysis is a diagnostic tool that makes networks in an organization visible and transparent, not a means in itself. Therefore any organization network analysis starts with a clear definition of the purpose of the network scan. For instance: is the purpose to identify unlocked innovative potential and grow an organization's innovation community, to evaluate the functioning of an organization's R&D department, or is it to identify knowledge leakage risks and informal blockades within the organization's innovation community that need to be addressed?

This need identification provides focus and serves as a point of departure for the next three steps: data collection, analysis and visualization, and inter-pretation. Combining network analysis of expertise with managerial and specialist knowledge of the organization provides the difference between a simple description and real impact of a network scan. An effective intervention always starts with a clear goal and objective, to which the corresponding networks to be identified and subsequent network measures to focus on are derived.

Need for transparency and commitment

Whatever the purpose of the network scan, it is key to ensure commitment by all in the organization to the purpose and set-up of the ONA, if only since without commitment, the data collected as input for the ONA is incomplete, rendering the data useless. This requires that goals and objectives must be clearly communicated to participants before, during, and after a network scan. Since there is a need for a very high response rate, nearly all individuals in (the relevant part of) the organization need to be committed enough to at least fill out the survey truthfully. With a small target population the data can be collected via face-to-face interviews. As the target population grows, this quickly becomes too labor-intensive, though. If one is to ensure commitment to the ONA when administering a survey, as a researcher one must be transparent about the purposes of the research and the way in which the analysis can be traced back to specific individuals. When identifying and communicating the main intention to conduct a network scan, getting all stakeholders involved early on in the process is key. The identification of the relevant stakeholders depends on the purpose of the network scan, yet in the context of innovation, typically encompasses commitment by TMT and key representatives of an organization's R&D and/or new business development department, representatives from key others in BUs involved, and representatives from the general labor force. While innovation commonly holds a positive connotation, an enhanced focus on innovation mostly is due to increasing pressures on the market, implies changes in the organization, and may therefore create resistance within the firm, reducing people's inclination to participate in the network scan. Commitment to the purpose of any network analysis project often requires that the confidentiality of data is guaranteed. Employees must feel that their privacy is protected. The closing paragraph of this chapter elaborates in more detail on this crucial aspect.

Step 2: Data collection

Data scoping. With the purpose of the network scan clearly identified and stakeholder management in place, we can move to the next step: the selection of the appropriate data collection method. We outline how to conduct a network assessment and the advantages and disadvantages of various methods available. Identification of the proper characteristics of the specific network scan to be conducted in an organization requires answers to the following six methodological scoping questions:

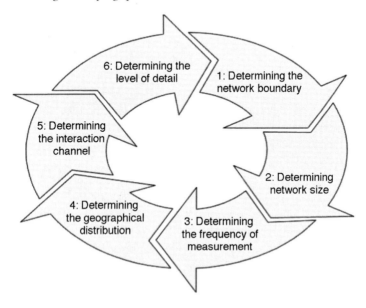

FIGURE 9.1 Data scoping

1 *Determining the network boundary:* Do we deal with a clearly defined target group, or not? A group of R&D scientists at a nutrition laboratory has clearly defined boundaries for the innovation community whereas the innovation network at a financial institution, or municipality does not have such a boundary. The way in which the innovation network must be measured differs between these two different circumstances.
2 *Determining network size:* Do we deal with a small or a large network? Are all respondents identified early on, or is the boundary of the network unclear at the start? A large network requires a different methodology than a small one.
3 *Determining the frequency of measurement:* Does one need a one-off network study, or is a longitudinal study with pre-and post-intervention measurements required?
4 *Determining the geographical distribution:* Are the targeted respondents geographically dispersed or not? Is one, or are more Business Units involved?
5 *Determining the interaction channel:* Do we deal with face-to-face or online interaction, or both? Is a technically mediated process of interaction newly established, or part of the existing organization routine/ procedure?
6 *Determining the level of detail:* Do we require general network information or is detailed network information necessary?

Depending on the answers on these questions, different network survey approaches and analysis techniques can be adopted. The answers to these questions help set the methodological scene, determine the scope and the approach to be taken, and result in the actual design of the data collection process.

Design of the data collection process

Case study. Studies using organization network analysis typically are, and cannot be anything else but, case studies (Yin 1994). A cross sectional research design is relatively meaningless for organization network analysis: it is conceivable that individuals from one organization studied could be in contact with other individuals, from another organization. Combining data about the organizational networks of different organizations in a cross-sectional research design demonstrates that these different individuals might well be able to link up. Organization network analysis allows for easy combination of qualitative research such as with interviews with quantitative research of network data. Indeed, combining the two would be hugely beneficial. One would even suggest that the many case studies performed in the social sciences should be accompanied by an organization network analysis, and that there should be no management intervention without a prior understanding of what the relevant social networks look like.

Data collection. Data collection is the cornerstone of any network related study. One should be very clear as to how social network data is collected, following social network survey methodology closely (Borgatti and Cross 2003; Wasserman and Faust 1994). One can either be in a position to define a relevant group of individuals in advance, or be forced to resort to a method by which one gradually learns what the relevant group is. The first is a sociocentric approach and the second an ego-centric approach.

In organization network analysis, the relationships are seen as links, or ties, between nodes that can be illustrated visually and mathematically (Whelan et al. 2011). So any assessment starts with plotting these nodes and their ties. In many cases a simple (online) questionnaire can help in quickly assessing the state of a network and in determining what is going on within the firm on a particular topic. To sample the network, two options are available: *socio-centric* and *ego-centric sampling*. Ego-centric sampling usually involves snowball sampling, rolling a survey out to new individuals who are mentioned in a previous round, until no new names are mentioned. An ego-centric approach is particularly suitable for those situations where mostly strong ties are present. The chances that relevant individuals who are not included in a first round will remain unmentioned is lower when ties in the relevant network measured are strong. Below each of these sampling procedures are highlighted.

- *Ego-centric sampling:* each actor/respondent lists their contacts to a particular topic, e.g. the discussion of new ideas and innovation of relevance to the firm. Names to be mentioned may be restricted to a particular number, may be suggestive of a number of names to be mentioned by a respondent, or may be entirely open.
- *Socio-centric sampling*: usually this involves the roster method where all names of the relevant community are provided in a list the respondents selects their contacts from (e.g. all BU members). This method relies strongly on instant recollection of the respondent. The framing of the name generator questions typically focuses on the most important relations maintained (e.g. who are the five most important persons with whom you talk about topic X).

When one is interested in a formal network in a firm, a socio-centric approach is feasible; when one is interested in a grouping that is not imposed but rather endogenously emerges over time, an ego-centric approach should probably be used (Wasserman and Faust 1994). This latter category is typical to the innovation network within an organization. When the boundaries of a network are not clear from the start, the snowball procedure is a commonly applied approach that helps to gradually uncover the boundaries of a network.

- *Snowball method*: When the target population is not clear from the beginning (Marsden 1990, 2002; Wasserman and Faust 1994), which is often the case for an innovation community, a snowball procedure should be considered as the most suitable approach of data collection. Snowball method involves several rounds of surveying or interviewing where the first round helps to determine who will be approached as a respondent in the second round, and so on. The first round of snowball sampling can be totally at random but it can also be based on specific criteria (Rogers and Kincaid 1981). To reduce the risk of ignoring "isolates", i.e. isolated persons within the organization who do possess relevant knowledge to a particular subject, but who may not be included in the study as their names are not mentioned in early rounds of surveying (Rogers and Kincaid 1981), the surveyor can opt in a first round to target respondents selected in collaboration with an organization's key managers such as the innovation management team. Conducting a number of interviews allows the researcher to become familiar with the organizational setting, to design a tailored-made network questionnaire that takes context specific terminology into account, and, foremost, helps to select key players. Starting with these key players, name generating questions such as the ones listed in Table 9.1, generate the names of a second wave of individuals, leading up to subsequent round of surveying, until the entire set of individuals with whom the innovative project teams maintain interaction, is uncovered, typically stretching across team boundaries and reaching many different units in the organization.

There are several aspects one should be cautious about exploring the innovation network when following an ego-centric approach: (1) not to miss out on isolates, and (2) to achieve network closure. The first one is achieved by combining a social network survey with interviews, at least with a number of key informants. The second one is achieved by properly snowballing the survey. Unlike a socio-centric study of, for instance, the complete email traffic within an organization, ego-centric studies that use snowballing in case of a potentially large sample size, face the problem that one should be

TABLE 9.1 An overview of common organization network types

Network	Name generator questions	Rationale	Sources
Formal (work interaction) network A network constituted by formal relations; "The prescribed roles and linkages between roles set forth in job descriptions and reporting relationships" (Aalbers 2012 p. 26).	"With whom do you interact to successfully carry out your daily activities within the organization that were prescribed or mandated by the organization? Who provides you with inputs to your job and to whom do you distribute outputs from your own work?"	The formal structure governs to a large extent who interacts with whom and it is this formal interaction that forms the basis for innovation.	(Mehra et al. 2001; see also Brass 1984; Brass and Burkhardt 1992; Cross and Cummings 2004; Whitbread et al. 2011)
Informal network Contacts useful in staying informally informed about what is going on within the organization.	"Who are the people that you connect with to discuss what is going on within the organization to get things done that are of personal relevance to you."		(Mehra et al. 2001); Smith-Doerr et al. 2004)
Innovation network Contacts are useful in helping you to be creative and innovative in your job beyond your daily work interaction routines, such as helping you to generate new ideas.	"Who are the people you connect with to formulate and discuss new ideas and innovations relevant to the company?"		(Rodan 2010; Cross and Prusak 2002; Aalbers et al. 2013, 2014)
Advice network Who goes to whom for advice on work-related issues?	To whom do you go for expert advice in doing your work?	Simply assessing who communicates with whom does not assure that the interactions present exchanges of information important to perform one's work. Especially in efforts that entail a collective to effectively pool its knowledge (e.g., new product development), it is essential to understand the effectiveness with which a group interchanges information.	(Cross et al., 2002 Cross and Cummings, 2004) Ibarra, 1993

TABLE 9.1 Continued.

Network	Name generator questions	Rationale	Sources
Customer knowledge network	With whom do you discuss customer needs and market demands?		(Krebs 1996)
Multiplex networks (rich ties)	No separate name generator question; a multiplex, rich tie is defined when formal and informal contacts between any two individuals exactly overlap. Other ties then are formal-only or informal-only.		(Aalbers et al. 2014; Ibarra 1993)

less sure as a researcher that network closure was reached. Typically, beyond an upper bound of some 300 individuals, individuals' recollection may not be relied on to provide all names of those involved in a particular network. In addition, in a larger group it can be more difficult to persuade enough people to participate in the data collection effort – the high response rate needed may be much more difficult to achieve. Approaching each actor individually may no longer be feasible.

Name generators

Organizational network questions are designed to obtain specific information about the relations people have with all the other members of an organization (Zagenczyk et al. 2011). Respondents have to scan their social contacts on factual (e.g. frequency of the contact) or attitudinal criteria (e.g. trust in someone) and determine with whom one maintains which contacts. Network name generator questions are typically used to collect the network data. The network questions that generate the networks need to be carefully chosen yet can be adjusted to the research setting so that all respondents (e.g. blue and white collar, technically skilled and managerially skilled) interpret them in the same way, are able to answer them and are willing to answer them properly (De Lange et al. 2004; Dillman 1978 [2000]). Regarding the respondent's willingness to answer, organizational network questions are different from standard survey questions in at least two ways. First, organizational network questions are often perceived as "sensitive" (Tourangeau et al. 2000) or "threatening" (Sudman and Bradburn 1982) by respondents. These questions can actually be viewed as intruding on an individuals privacy, and respondents may be anxious that their answers will be exposed to people other than the researcher (Tourangeau and Smith 1996; De Lange et al. 2004).

What name generator question to use is key (Borgatti and Cross 2003; Rogers and Kincaid 1981; Rodan 2010; Contractor et al. 2011). The formal (mandated) or work interaction network, for instance, is measured by asking respondents to indicate the persons with whom they exchange information, knowledge, documents, schemes and other information sources to successfully carry out their mandated daily activities (Mehra et al. 2001; Krackhardt and Hanson 1993; Ibarra and Andrews 1993; Burt 1992; Cross et al. 2002). The emergent, informal network is measured by asking with whom one chooses to informally discuss what is going on within the organization – this is sometimes referred to as "the grapevine" (Ibarra 1993; Mehra et al. 2001). The question to determine the innovative knowledge transfer network asks individuals about those with whom they exchange new ideas, innovations and substantial improvements to products and services that are not part of their day-to-day activities (Rodan 2010; Cross and Prusak 2002; Aalbers et al. 2013). Whereas the name generator question for the formal network measures the connections resulting from exchange of routine issues and day-to-day information, the name generator question for the innovative knowledge transfer network asks about the transfer of new or complex knowledge that was specifically not perceived as related to the ongoing business of the organization (Aalbers et al. 2013; Harrisson and Laberge 2002). Table 9.1 summarizes some of the most widely used name generator questions, and highlights the rationale for choosing one over the other.

Other data

Organization network analysis researchers have a tendency to ignore how differences between individuals might affect organizational network usage and benefits (Anderson 2008; Aalbers et al. 2013). Nonetheless personality characteristics have occasionally been linked to network position (i.e. Burt et al. 1998; Oh and Kilduff 2008; Foss et al. 2009). Moreno (1961) and Simmel (1950) already emphasized the relevance of linking social structures and psychological processes. In this chapter on network methodology we therefore stress the relevance of collecting additional data at the individual, but also the team level to provide input to more accurately understand what drives particular network configurations. For instance, management may want to explore whether individuals with certain predispositions are indeed better connected than others in a knowledge transfer network, for instance in terms of closeness centrality, or more engaged in inter-unit knowledge transfer (c.f., however, Aalbers et al. 2013).

Designing the questionnaire

Collecting data within an organization is typically conducted by means of an online questionnaire, containing questions as outlined in Table 9.1 to identify the network position of individuals as they contribute to their own or their team performance (Marsden 2002). These questionnaires can be expanded to include additional questions on employee characteristics, behaviors or prior experiences to allow for even richer data analysis.

The high sensitivity of organizational network questions can lead to (item) non-response rates when compared to more generic questionnaires. At the same time, non-response is much more problematic when doing organization network analysis. Ways to counter this risk and the subsequent bias in the data that is collected, include precise, concise and simple formulation of the question in the questionnaire, prior to its distribution. To reduce ambiguity regarding the interpretation of the questions by the respondents, network questions can be formulated in the native language. The sensitivity of these questions can be softened by choosing more neutral wording, dependent on the context (Dillman 1978 [2000]). The survey questions should also be written in such a way that makes it uncomplicated for the respondents to answer (De Lange et al. 2004). Keeping the number of survey questions to a minimum avoids distraction from the objectives defined under Step 1. Response can be enhanced when it is signaled that the organization and its management support the study, evident for instance from a personalized cover letter or email, signed by a senior sponsor such as an innovation director or member of the board. After distribution of the questionnaire, close monitoring and swift follow-up towards the target respondents is vital. Respondents who do not reply initially, may for instance be approached personally to fill out the questionnaire.

Securing an accurate measure: survey response rate

It is vital to achieve a response rate of at least 90 percent when performing ONA. Executive support has been found to make a difference in gaining commitment among prospective respondents. In addition, respondents should be well-informed regarding the intentions, purpose, and possibility for anonymity of the study. In our studies, typically a senior executive sends out an email to the identified employees preparing them for the survey. In this message the executive requests that every recipient is to participate and ensures that individual responses will be treated confidentially. Only aggregated or anonymized results are shared within the organization. A response rate below 90 might lead to an incomplete network being mapped, which in turn can lead to misinterpretation of results and respondents' positions in the network.

Mapping networks of social interaction relies strongly on the deployment of name generators, questions to find out what social relations individuals maintain. Name generator questions, may be highly suggestive of the number of names to be provided, or leave the number of names open. Each approach may introduce a bias.

> The first might make some respondents with a limited number of contacts list contacts they have tenuous contact with, and might make those with many contacts list only their most important ones. The latter, the free-recall method, relies on respondents' memory but is suggested to be most suitable in a study where network boundaries cannot be determined a priori.
> (Aalbers, Dolfsma, and Koppius 2013: 9; Friedman and Podolny 1993)

Step 3: Analysis and visualization

If data of high quality is collected the next, and for many, the defining step in the analysis is to visualize the networks. There are a number of different software programs available for this, ready for use, and freely available, often that only require input of data in the right format. We have used several different programs for our own research, including Netdraw, available in UCINET, NODEXL, and Pajek. With network visualizations becoming common in science in practice, various techniques are available to make a good visualization of a large network by using "R" as well, as an open source free software programming language and software environment. The latter is especially appropriate when encountering substantial datasets, for instance in the case of analyzing longitudinal and/or multilevel data. Each of these options have their own manuals, and online forums readily accessible for the necessary introduction.

These software programs for analysis of social network data, as well as the advanced statistical techniques they make use of, continue to improve. However, discussing the current state of the art will be outdated soon.

A number of considerations should be kept in mind when doing visual or statistical analysis of network data. One needs to remember the purpose of the analysis, from both an organizational as well as an academic point of view. There is a plethora of measures available, more than we discuss here, as well as varieties of the measures that we do discuss in these pages. Providing descriptive summary statistics is helpful, but mostly for relevant measures: an understanding of the organizational context and the goal of a study is indispensable when determining which measures are relevant. Relevant measures for an understanding of a network include network density, network cohesion, clique formation, lambda sets.

An almost indispensable step to determine which measures are relevant again is a visualization of the networks. Inspecting a picture of a network often is the be all and end all of ONA, but a network figure can be made more informative by further characterizing both the node as well as the ties between the nodes. One should be aware, however, that network figures can and often are purposefully presented in the way that they are. In particular, the vicinity of nodes to other nodes is something that can be manipulated. Bar manipulation of the data itself, presence (absence) of a node and its ties in a picture cannot be manipulated, but positioning can. For purposes of clarity and understandability of a visualization, most programs allow for the option of introducing thresholds for inclusion of a node in a figure. Nodes with ties that do not meet certain criteria, such as a minimum number of ties they hold, can be left out of a picture. It is important to realize this when studying what might otherwise appear to be an intriguing figure – a network figure can suggest more or different things than it actually shows. In addition, networks are dynamic over time: any figure is a snapshot at a particular point in time. Given the right data, the development of a network can be presented as a "movie". The development of a network can also be simulated using such programs as SIENA.

Once one has analyzed the shape and structure of an organization's social networks, depending on the purposes one has in mind, a plurality of sophisticated quantitative analyses may be conducted on the data. Detailing these more is beyond the scope of this book. Please refer to Wasserman and Faust (1994), among others, for more details on quantitative ONA.

Step 4: Interpretation

With data collection and the subsequent analysis and visualization behind us, the interpretation phase is the moment to "connect the network dots" to the goal and objectives behind the network exercise. This phase is important as it provides the difference between "data" and "impact". The goal and objectives established in the need identification round are revisited and those network characteristics relevant to the particular needs are highlighted and assessed.

A hypothesis defined in the need identification phase, such as "this organization lacks sufficient knowledge brokers to direct novel ideas to the relevant pockets of functional expertise" or, "Department A is the spinal column to any successful new business development project," are tested on the corresponding network characteristics. For instance the number of brokerage positions occupied by members within the organization, or by Department A in relation to other departments is established to see if these amount to what one would expect to find. The stability of these positions can be tracked over time. The same may be done for overall network density, for instance after an intervention was put in place to steer towards the needs identified during Step 1.

Contrasting these observations and initial findings with previous experience within the firm is next. In some cases, prior surveys on employee satisfaction, team performance, or Business Unit profitability may provide added information to point to reasons for structural differences in parts of a network. Organizations that have conducted network assessments before can consider, if properly performed, to do a dynamic analysis using both the datasets.

At the very least, if the data is of insufficient quality, or if there has been too much attrition, benchmarking prior network patterns to more recent patterns can be undertaken. In some cases a comparison to observation at other (partner) firms may be a possibility. The outcome of the network observations in combination with a "reality check" regarding what is currently going on within the firm results in a set of actionable insights.

Post-assessment interviews

Getting the most out of network data analysis requires in depth analysis of the findings in a way that takes the particular situation of the organization (or BU, or project) into account. Interviews with a select number of employees helps to better understand the dynamics behind the network. Network analysis typically is multi-method. Depending on the goals and objectives identified under Step 1, both individuals can be identified whose positioning appears to be close to, or, instead, is far removed from, the network characteristics envisioned. Consideration of centrality scores, brokerage scores, and the fulfillment of multiplex roles in relevant networks can foster helpful dialogue about the shape of parts of a network and the possibilities for improvement. Maintaining confidentiality, such dialogues can be conducted successfully interpreting the aggregated results from a network analysis.

Prioritization

Prioritization of actions based on current business objectives is the next step. This typically happens in parallel with the design of additional interventions that help to build towards the objectives set out. Network analysis is helpful in tracking the progress made. Tracking of unutilized knowledge and expertise can and should be an ongoing process – an understanding of potential opportunities for better collaboration within an organization are not captured by a single snapshot of the firm.

Monitoring

As for any change trajectory, intervention targeted at improving innovation performance of an organization requires close monitoring. Such monitoring begins with the definition of metrics to measure the performance. Specific connectivity and business metrics may be defined to be assessed in the months following the initial network assessment. Depending on the type of network intervention a period of 2 to 6 months is realistic to expect actual progress in terms of appropriately altered social relations in the organization. Just as in the prior steps, transparency remains central throughout this process. Results and next steps should be communicated to the stakeholders identified as early and openly as possible. Based on our experience with interpreting network based findings within a firm, having a variety of stakeholders on board when interpreting the findings and determining subsequent priorities is to be attained if only because it leads to commitment to ongoing data collection and collaboration with change processes in an organization.

Final reflections on network data collection

Network data collection is no mean feat. The need for high response rates and privacy considerations make network analysis a managerial tool that requires prudence and thorough preparation. Partial data collection is not worth much and hampers willingness among stakeholders (employees and management) to engage with such a procedure in the future. At the same time, however, an organizational setting provides numerous options to collect network data in a systematic and transparent manner, even for part of the organization, as long as the boundaries of the relevant social network are well-defined.

New technologies, including the adaption of Enterprise Social Software (ESS), provide opportunities to map organization networks on specific innovation related topics. ESS includes corporate intranets and other software platforms aimed at enhancing productivity, improving communication, and promoting collaboration. Examples of ESS include platforms such as Jive, IBM Connections, Microsoft (including Yammer), Salesforce.com (including Chatter), IBM's Connections as well as many company-proprietary platforms. These platforms make it possible to follow people across divisions and hierarchies over time. But also less tech-savvy data collection opportunities exist, that to date, have remained largely underexplored. For instance, project evaluations and year-end feedback forms could provide structural input on collaboration patterns within a firm. A main drawback to these sources is the embeddedness of these modes in the daily operations of most organizations. As such data is collected ex-post and for purposes different from that of organizational network analysis, ensuring transparency and openness can sometimes be more difficult. Embedding of the data collection procedure in an organization's day-to-day processes can, however, be a major support in capturing the complete network of interest, and in inserting network analysis into a mainstream tool for reflection on operational and innovative performance.

Research ethics in organization network analysis

Ethical considerations are part and parcel of any organizational network analysis exercise. It is important, under all circumstances, to be considerate of respondents' privacy and rights, the company's desire for anonymity and responsible behavior, and the possible implications of the study's results for the company, its employees, and other parties. Each study should adhere to the notion of informed consent: "participants' right to be informed about the nature of a research study and its risks and benefits to them prior to consenting to participation" (Hesse-Biber and Leavy 2010: 61). An important step in this process is to provide as much information as possible about the network analysis to be conducted with regard to the intentions and outcomes. The privacy of individual respondents is typically, and relatively, easily safeguarded by anonymizing, only analyzing at a high level of aggregation or disclosing results to a select group of individuals only. Respondents should be provided with the name and contact information of the researcher so that they may be contacted when respondents are feeling dissatisfied or have questions.

Network data itself is insufficient to establish if an individual, team or unit performs well. Additional information is usually required for that. Observations such as: "as head of the Marketing department, Martha clearly has too few connections with the rest of the business to stay in tune with what is going on within the firm", or "Dirk is clearly not central enough, as he falls outside the epicenter of our organization's network", are likely to be shallow. We would like to warn against quick and therefore superficial and possibly misleading conclusions based on ONA. Departments, teams and individuals may, of course, be connected better (worse) than one would expect given the goals and position they have, but drawing conclusions based on a single measurement, and without taking other non-network information into account, is very likely to be superficial at best. Networks are dynamic by nature, develop over time, and relations between individuals take shape at various levels and in different types of networks concurrently. Simple network logic also points out that individuals cannot all be highly central at the same time.

10
MANAGING INNOVATION IN THE NETWORKED ORGANIZATION

Conclusions

People who collaborate bring innovation in an organization to life; a view of the organization as an inherently networked entity is best suited to understand and manage this.

Firms are thoroughly social entities – it is because of the benefits of joining forces that firms exist in the first place. Joint production, also of the innovative variety, necessarily involves the exchange of knowledge, in some form, of some kind. Exchange of knowledge, or communication generally, even in the smallest of organizations soon becomes highly complex. This book offers a number of powerful tools to visualize and understand the communication and exchange structures in an organization. The tools we present are available in the social sciences generally, and organization network analysis in particular. In our own work we have both sharpened the tools allowing for more detailed analysis of an organizational setting, and we have also made it more accessible for a broader and managerial audience. The latter we do, for instance, with this book. In doing so, we have tried not to lose the sharp edge that network analysis can provide.

We offer a number of suggestions throughout this book, drawing on insights from social network theory, sharpened by research in a number of different empirical settings that we have undertaken by ourselves. Where possible, we were clear about the specific organizations these are, such as in the case of Siemens. We are not in a position to mention all of the names of the organizations that we have researched and consulted for. The firms are diverse in their characterization, and include firms active in engineering, financial services, consulting, manufacturing, food processing, and R&D/hi-tech organizations.

> Horizontal and vertical network relations, crossing unit and hierarchical boundaries, determine individual and team success.

We purposefully alternate between offering real-life (business) examples and more rigorous analytical approach, believing that both these can strengthen each other. A manager in an organization who uses insights we offer without having at least an understanding of the academic background drawn upon, may be enticed to consider the wrong interventions. An academic not concerned with real-life settings will pursue lines of research that are irrelevant for society and will, sooner rather than later, also prove to be an academic cul-de-sac.

> Management needs to recognize the central and fundamental importance of effective communications and interpersonal relationships for an organization to successfully innovate.

We believe that the emphasis on communication and knowledge exchange in an organization could give rise to a new function of the formal or informal variety: the communication director. As knowledge develops quicker, and as knowledge newly to be developed draws on diverse knowledge sources, proper communication is vital, and a role for a communication director is evident. Those who are innovation managers by function, in part actually are communication directors in all but name if one considers their activities. Using the ideas and notions in this book may help them become even more effective in their role. The same thing holds true for Chief Information Officers (CIO's) who should focus less on the knowledge items a firm has, but on the routes they do or should take flowing through an organization. CIOs should also be much more aware than they are now about flows of knowledge beyond the formal connections, and beyond the use of IT in support of communication.

> Managers can orchestrate innovative knowledge flowing through their organizations by identifying and directing the formal and informal brokers, brokering ideas, in the different phases of innovation process.

While we have mostly focused on how a better understanding of communication patterns in an organization can make the organization more innovative, communication also undergirds other activities of and outcomes for the organization. The insights we offer can thus, with some alteration, be used to promote those other activities as well. Communication and exchange being so central to innovation, however, a focus on innovation will allow for a clear exposition. In addition, of course, innovation is a core activity for firms.

> Multiplex, layered relations among employees offer opportunities to develop new and enduring network relations beneficial to innovation.

For management and employees alike, it is important to realize the existence of the multiple pathways that exist for the transfer of innovative knowledge. For innovative knowledge to most successfully transfer, we find that layered relations among employees offer opportunities to develop new and enduring network relations beneficial to innovation. We outlined that the multiplex combination of a formal tie and an informal tie contributes to knowledge transfer beyond the effect of either in isolation. Such multiplex relations are found to have a particularly strong effect on innovative knowledge transfer in an organization.

> Downsizing and other major organizational changes need not be detrimental to innovation when ex-ante network positions of employees are taken into account and downsizing is executed such that key positions are protected and strengthened where needed.

In various chapters we have pointed out the importance of diversity for innovation. Managers can preside over major improvements in the conversion of external knowledge into innovative outcomes, when they understand the relevance of sufficient brokerage activity to foster diversity. Each chapter however, illustrated in various ways that generating diversity and sparking new ideas as a means to foster innovation must be viewed upon as a social process that spans across multiple employees.

The intra-organizational network methodology presented in this book in the form of the innovation engagement scan (IES), can be relatively easily adopted in itself or as part of a larger effort, and might equip management with the means to monitor the effect of their own or external managerial actions on the innovative activity within the organization. By using the intra-organizational network methods, as displayed in various forms throughout this book, managers gain a bird's-eye view of existing network structures and communication patterns that facilitate the innovative activity within the organization. This might raise awareness of potential risks with regard to the innovative capacity, such as dependencies or underutilized potential, which could be input for more directed managerial action.

We showed the positive effects that newcomers might have on the innovation network in response to directed managerial intervention. Managers should be aware of the extended effect of an individual's position on the overall network. Thus they might want to specifically focus or even invest in these dominant actors and the relationships between management and these actors to reduce the risk of bottlenecks and ensure an efficient flow of innovative knowledge. Directed intervention under conditions of organization turmoil, in turn, we showed may benefit from a network point of view prior to such an event to take into account the ex-ante network positions of individuals. Insights rendered may help management to carry out a reorganization in such a way that key positions relevant for an innovative social infrastructure are protected and strengthened where needed, within the realistic (and legal) constraints that come with any drastic restructuring of a firm.

Drawing on our discussion, we would like to offer additional recommendations in Table 10.1 mostly for managers who seek to enhance their organization's innovativeness. The insights, we suggest, will also offer suggestions for fellow scholars in terms of future research directions.

TABLE 10.1 Key network take-aways for innovation

Characteristics of:	An organization network conducive to innovation	An organization network shy to innovation
Orientation	Open; external (other units, outside of firm).	Closed; Internal.
Bridges to different others (structural holes)	Many: their contribution appreciated and supported.	Few; seen as causing delays, inefficiencies, and 'costs'.
Multiplexity	Layered, overlapping and mutually supportive networks, combining formal and informal relations, actively used to promote innovation.	Low use of multiplex relations, or use of multiplex relations to prevent change and hamper innovation.
Managerial directive	Low, but targeted.	High, piece-meal.
Roles/position	Diversity of roles, actively monitored for their contribution to strategic innovation goals.	Role diversity low; not monitored for contribution to strategic goals not facilitated.
Benefits	Fast access to heterogeneous information, enhanced creativity, more innovation. High adaptivity, ready for external shocks.	Efficient communication, shared norms and high trust.
Drawbacks	Inefficiencies due to redundant networks, and additional communication costs.	Redundancy in communication, low chance of unexpected and creative recombination of knowledge.

APPENDIX

Supporting notes to Intermezzo Case "Cooperation for innovation at Siemens"

These notes are structured such that they correspond with the questions listed at the end of the case. The answers and suggestions provided in these notes are by no means "definitive" but intended to fuel discussion on the topic of organization network configuration, individual and group positioning within these networks, and the rationale and workings behind a network intervention.

1. **Which network do you think influences the innovative capacity of Siemens Netherlands most: the formal or the informal network? Why?**
   Several networks can be distinguished. A widely used distinction is that between the formal and the informal network. The formal network is the pattern of formal relations within an organization. Even though the informal network is often believed to be more important for innovation, at least for SNL the formal network seems to be slightly more important. This might be related to the need for formal commitment (of resources) during the innovation process.

2. **What are the main benefits of the formal network?**
   Benefits of the formal network comes from its relative transparency. Due to the fact that the formal structure is based on the formal hierarchy people seem to be aware of who is supposed to interact with

whom. According to Adler and Borys (1996) the formalization of communication within an organization reduces the chances of conflicts occuring. What is more, the formal structures reduce possible ambiguity about the way an individual employee is expected to perform, which in turn increases the individuals appreciation of the work that is conducted.

3   **What are the main benefits of the informal network?**
Benefits of the informal network come from its flexibility, both in using the network and in allowing it to be changed by the individuals themselves.

4   **How to influence the formal network?**
In the literature there are several authors who stress the correlation between the formal and the informal network. The informal network can supplement the formal network to make sure information flows between units that are not formally connected with each other. When altering the formal structure of an organization to adjust to changing (market) circumstances it can be useful to draw from existing relations from the informal network. One could thus formalize informal relations, or give persons occupying central positions (nodes) within the informal network a role in the formal network. This would allow him/her to leverage the informal network.

5   **In which way do both networks clash and in which way do they supplement each other?**
Existing relationships in the formal network can be adapted based on the informal network, building on levels of trust between the individuals involved that have already evolved.

One possible negative effect due to the co-existence of a formal and an informal network: the informal network can undermine the exciting authoritarian relations of the formal network. This drawback triggers the discussion on professional versus informal authority.

6   **In all the networks many relations seem to be one directional. This implies a low reciprocity. What could be the consequences of this observation?**
The fact that communication and knowledge transfer is predominantly one-way indicates that knowledge sharing within SNL could be

improved. It is known, however, that individuals in central/powerful positions tend to receive more (information) than those who are more peripheral. On the other hand the lack of reciprocity can also result from a relative discrepancy in level of expertise on certain knowledge domains. In this case low reciprocity is a result of knowledge based authority, due to which a relatively large number of employees turns to the expert without reciprocating the effort by providing information in return. A result of low reciprocity can be a lack of trust within the relationships (e.g. Nooteboom 2002) and in the end within the Siemens organization. Due to this latter risk, a certain degree of reciprocity must be present. Approaching the low reciprocity from a more optimistic point of view, the one directional communication implies a certain degree of openness within the communication structure. Individuals are (until so far at least) willing to provide information to peers without putting their own interest first. Given human nature, it is questionable if this change in behavior will be permanent.

7   What do you think about the position of Hans, the Senior Innovation Manager, within the network? Should he strengthen his role as a link pin between the different divisions or should he delegate this role more to employees on lower levels in the organization?
    Hans should be aware of his role as catalyst. He should not strive for a permanent position in the communication networks related to the theme of transport. However, he must be aware of the crucial role he now plays within the networks – turning his back on them would jeopardize them. He will thus have to phase out. His influence on the way in which the formal network is structured is much bigger and more directly discernible than his influence on the informal network. Nevertheless, he must keep an eye on the way in which the informal network might influence the effectiveness of the formal network.

8   What could Hans, the senior innovation manager at Siemens, do to improve inter-divisional communication, based on the communication structures described above?
    When discussing the possible improvements of inter divisional communication it is important to note that communication is based

on the social interactions of individuals. Some suggestions as an answer to further improvements of the network are the following:

- Create sufficient time for the knowledge sharing to occur.
- Facilitate contacts that are not evident.
- Create a climate that nurtures experimentation.
- Reduce the dependency on a small number of persons within the networks.
- Take into account the motivation of the individual employee.

9 **Which approach would be more helpful to Siemens: using information about centrality to determine ones communication profile or using information about brokerage? And why?**

This question is meant to trigger the discussion of two different approaches in selecting persons highly involved in the communication process within an organization. Centrality identifies persons with a high profile within the organization. These persons are highly involved in transferring information and are therefore highly aware of what is going on within the organization. Centrality is therefore an indication of the degree to which individuals have access to resources like knowledge (Hoang and Antoncic 2003). Using centrality is therefore a useful approach in studying the diffusion of innovation within a company (Ibarra and Andrews 1993). The benefit of approaching knowledge transfer from a brokerage point of view is that brokerage includes information regarding the direct interaction with other people, but contrary to centrality, thereby taking group membership into account. In the case of Siemens one could therefore argue that brokerage generates more detailed insight in the degree to which knowledge is crossing divisional borders. This is important as diversity of knowledge resources to a large extent determines the degree to which the creation of new ideas and inventions is fostered (Granovetter 1973; Nonaka and Teece 2001; Schulz 2003).

10 **In which way could the IM department apply the gathered information about: a) centrality and b) brokerage in a practical manner, to improve the innovative climate within Siemens regarding transportation?**

The short answer is by counting the number of contacts (formal and informal) each member of a network entertains. Such counts can then be used to analyse his/her communication pattern. Interesting results can emerge, where particular individuals might turn out to be more important for knowledge transfer with the company than would be predicted on the basis of this person's formal position.

When observing the communication patters at an individual level, it is also possible to deduce valuable information about the position of the individual within the group as a whole, and the different sub-groups. A useful indicator when discussing the position of individuals within a network is that of *centrality*.

Centrality gives an indication of the influence of an individual within a network with regard to all other individuals within that network (Brass and Burkhardt 1992).

There are three measures for centrality (degree, closeness, betweenness) that each provide an evaluation of the locations of individuals within the network in terms of how close they are to the "center" of the action in a network. (For a comprehensive overview of these different forms and the advantages and disadvantages of each form see e.g. Wasserman and Faust 1994).

The definitions of what it means for an individual to be at the center differ. Determining centrality within a network seems abstract but in fact it is quite a simple exercise. When explaining the position of individual employees in a network it is useful to refer to the following three basic, ideal typical networks distinguished by Freeman (1979): the "star," "line," and "circle."

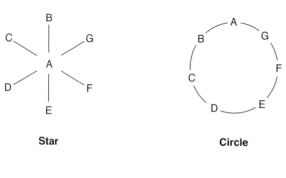

Looking at the star network it soon comes to mind that the position of person A will be most influential when exchanging information. Can a similar observation be made about the two other ideal typical networks? And for combinations of them? What is the advantage of one structural position over another? In the following section the position of person A will be discussed to explain the three different centrality measures that can be used to explain this issue of positioning.

Degree: Looking at the star network it shows that the position of A is advantageous because A has the opportunity to draw from alternative channels when the requested or sought after information is not provided by one of the other persons in the network. In the case that E decides not to provide A with information A can easily turn to C or F as an alternative. E does not have that option. When A, for instance, decides not to provide E with information then no information will reach E at all. It is for this reason that the dependency of E is far higher than that of A, resulting in a more powerful position for A. Degree centrality refers to the number of contacts an individual has within a network. In the star network, person A has a score of 6 whereas all the other actors have a score of 1. From a degree centrality point of view all actors in the circle network have the same number of relations (2). This means that all positions have the same advantages and disadvantages with regard to the access to information. In the line network it is the persons on the extremes, A and G, who have a structural disadvantage due to their position. They are highly dependent on one single person. Generally, though, actors that are more central to the structure, in the sense of having higher degree or more connections, tend to have favored positions, and hence more power. A different measure is needed to bring this out: Closeness.

Closeness: Another important reason why person A can be perceived to be of high influence in the star network is that person A is closer to more actors than any of the other actors.

Besides the influence an employee can derive from being in direct contact with a large number of colleagues.

In the star network this means that person A forms the center of attention. His views are heard by a larger numbers of actors. Actors who are able to reach other actors through shorter path lengths, or who can be more easily reached (shorter path lengths) by other actors, have favored positions (Borgatti 2005; Smith et al. 2014). This position forms structural advantage compared to the other actors in the network. Person A has a "geodesic distance" (average distance to other persons in the network) of 1 whereas all other persons have a higher score.

Now consider the circle network in terms of actor closeness. Each actor lies at different path lengths from the other actors, but all actors have identical distributions of closeness, and again would appear to be equal in terms of their structural positions. In the line network, the middle actor (D) is closer to all other actors than are the set C, E, the set B, F, and the set A,G. Again, the actors at the ends of the line, or at the periphery, are at a disadvantage.

11 **What are the strengths and the weaknesses of the five brokerage roles as described by Gould and Fernandez (1989) in Chapter 3? Which of these roles seem to dominate in the social networks of Siemens Netherlands?**
Besides using centrality measures as a tool to discuss the structure of both networks it is possible to look at the network positions of the individual employees from a brokerage point of view. The *direction* in which knowledge flows is explicitly taken into account. When looking at the social relations in more detail several sub-groups can be distinguished. Within the social network literature these groups are called clusters.

This distinction of communication roles is based on the "membership" of a certain division and the direction of the communication of employees. The different brokerage roles serve different purposes. Some are focused on what is going on within the unit (coordinator, gatekeeper), while others are more oriented to the outside (representative, itinerant broker, liaison). While there should be communication between divisions, not everybody would need to be equally involved in that. In addition, it is conceivable that a division can be focused too much on communication with the other divisions, or that it is focused too much on communication within its own division. One could, based on these considerations, determine an ideal communications profile for each (type of) position within a division.

12 **How would information about the centrality of individuals on the one hand, and information about brokerage roles adopted by individuals on the other hand be important to determine communication patterns within SNL? Why?**
Structural holes and non-redundancy: According to the literature looking at the link between creativity and network position it is important for innovation to occur to create an organization network that has

an open structure (Burt 1997). This as opposed to a clique network that is characterized by dense relations and a high overlap in relations between different groups in an open or entrepreneurial network. Such a general observation can be translated into a discussion about centrality of individuals in a network, and the roles they adopt.

Relations with individuals outside a cluster might mean for any given individual that he may not be as centrally placed within his own cluster (division). As one may not be able to maintain more than a certain number of ties, making sure that structural holes exist in each division, might put an additional burden in terms of internal communication on other individuals. Rather than having brokerage roles uniformly adopted within a division, one may want to make sure that certain roles are aimed at certain individuals.

When striving for innovation it is often argued that organizations should strive for the reduction of redundancy within the network. Redundancy refers to the extent of the contacts of a network. However, reducing redundancy within a network increases the dependence of that network on a few (or even a single) individuals' contacts with external parties.

13 Which roles seem to be most important in the early stages of an innovation project (i.e. idea generation etc.)?

In the beginning of the creative trajectory it is important to generate sufficient ideas and to stimulate creativity. A main factor in accomplishing this goal is sufficient diversity in resources. Hargadon (2002) refers to this process by stressing the importance of different knowledge domains when striving for innovation. In the discussion concerning brokerage roles this means that especially the role of liaison and representative are important to the externally oriented roles. Access to novel resources asks for a broker who draws from other divisions with other problems and perspectives on potentially similar problems or opportunities.

14 Does this role change when the innovation stage becomes more mature, and if so, why (not)?

Roles focusing more on communication with individuals outside of the division seem to be preferable in the early phases of an innovation project (Ancona and Caldwell 1997). In later phases, one should focus more on meeting goals set efficiently; therefore communication should be more internally oriented.

REFERENCES

Aalbers, H.L. (2012). *Organizing intra-organizational networks for innovation* (Doctoral dissertation, University of Groningen, The Netherlands).

Aalbers, H.L. and Dolfsma, W.A. (2008). Social networks: Structure and content. In J.B. Davis and W. Dolfsma (eds). *Companion to Social Economics*, Cheltenham, England: Edward Elgar, pp. 390–405.

Aalbers, H.L. and Dolfsma, W.A. (2014). Innovation despite reorganization. *Journal of Business Strategy*, 35(3): 18–25.

Aalbers, H.L., Dolfsma, W.A. and Koppius, O. (2013). Individual connectedness in innovation networks: On the role of individual motivation. *Research Policy*, 42(3): 624–634.

Aalbers, H.L., Dolfsma. W.A. and Koppius, O. (2014). Rich ties and innovative knowledge transfer within firms. *British Journal of Management*, 25: 833–848.

Adler, P.S. and Borys B. (1996). Two types of bureaucracy: Enabling and coercive. *Administrative Science Quarterly*, 41: 61–89.

Adler, P.S. and S. Kwon (2002). Social capital: Prospects for a new concept, *Academy of Management Review* 27(1): 17–40.

Agarwal, R. and Helfat, C.E. (2009). Strategic renewal of organizations. *Organization Science*, 20(2): 281–293.

Agneessens, F. and Skvoretz, J. (2012). Group differences in reciprocity, multiplexity, and exchange: Measures and applications, *Quality & Quantity*, 46: 1523–1545.

Ahuja, G. (2000). Collaboration networks, structural holes and innovation: A longitudinal study. *Administrative Science Quarterly*, 45: 425–455.

Ahuja, G. and Katila, R. (2004). Where do resources come from? The role of idiosyncratic situations, *Strategic Management Journal*, 25(8–9): 887–907.

Ahuja, G., Soda, G. and Zaheer, A. (2012). The genesis and dynamics of organizational networks. *Organization Science*, 23(2): 434–48.

References

Aiken, M. and Hage, J. (1968). Organizational interdependence and intra-organizational structure. *American Sociological Review,* 3: 912–931.

Albrecht, T.L. and Hall, B.J. (1991). Facilitating talk about new ideas: The role of personal relationships in organizational innovation. *Communication Monographs,* 58: 273–288.

Albrecht, T.L. and Ropp, V.A. (1984). Communicating about innovation in networks of three U.S. organizations. *Journal of Communication,* 34(3): 78–91.

Alcacer, J. and Zhao, M. (2012). Local R&D strategies and multi-location firms: The role of internal linkages. *Management Science* 58(4): 734–753.

Allen, T.J. (1971). Communications, technology transfer, and the role of technical gatekeeper. *R&D Management,* 1: 14-21.

Allen, T.J. (1977). *Managing the flow of Technology: Technology Transfer and the Dissemination of Technological Information within the R&D Organization.* Cambridge, MA: MIT Press.

Allen, T.J. and Cohen, S.I. (1969). Information flows in research and development laboratories. *Administrative Science Quarterly,* 14: 12–19.

Ancona, D. (1990). Outward bound: Strategies for team survival in an organization. *Academy of Management Journal,* 33: 334–365.

Ancona, D. and Caldwell, D. (1992a). Bridging the boundary: External activity and performance in organizational teams. *Administrative Science Quarterly,* 37: 634–665.

Ancona, D. and Caldwell, D. (1992b). Demography and design: Predictors of new product team performance. *Organization Science,* 3: 321–341.

Ancona, D. and Caldwell, D. (1997) Making teamwork work: Boundary management in product sevelopment reams. In M.L. Tushman and P. Anderson (eds). *Managing Strategic Innovation and Change, A Collection of Readings,* New York: Oxford University Press, 1997.

Anderson, M.H. (2008). Social networks and the cognitive motivation to realize network opportunities: a study of managers' information gathering behaviors. *Journal of Organizational Behavior* 29: 51–78.

Anderson, N., Potočnik, K. and Zhou, J. (2014). Innovation and creativity in organizations a state-of-the-science review, prospective commentary, and guiding Framework. *Journal of Management,* 40(5): 1297–1333.

Appleyard, M. (1996). How does knowledge flow? Interfirm patterns in the semiconductor industry. *Strategic Management Journal,* 17: 137–154.

Argote, L. (1999). *Organizational Learning: Creating, Retaining, and Transferring Knowledge.* Boston, MA: Kluwer.

Athanassiades, J.C. (1973). The distortion of upward communication in hierarchical organizations. *Academy of Management Journal,* 16: 207–226.

Atuahene-Gima, K. and Evangelista, F. (2000). Cross-functional influence in new product development: An exploratory study of marketing and R&D perspectives. *Management Science,* 46: 1269–1284.

Baer, M., Leenders, R.Th.A.J. Oldham, G. and Vadeera, A. (2010). Win or lose the battle for creativity: The power and perils of intergroup competition. *Academy of Management Journal,* 53: 827–845.

Balkundi, P., Kilduff, M., Barsness, Z. and Michael, J.H. (2007). Demographic ante-

cedents and performance consequences of structural holes in work teams. *Journal of Organizational Behavior,* 28: 241–260.
Ballinger, G., Craig, E., Cross, R. and Gray, P. (2011). A stitch in time saves nine: Leveraging networks to reduce costs of turnover. *California Management Review,* 53: 111–133.
Bandura, A. (1986). *Social Foundation of Thought and Action.* Englewood Cliffs, NJ: Prentice-Hall.
Bartlett, C.A. and Ghoshal, S. (2002). Building competitive advantage through people, *MIT Sloan Management Review,* 43(2): 32–74.
Bartunek, J., Balogun, J. and Do, B. (2011). Considering planned change anew: stretching large group interventions strategically, emotionally, and meaningfully, *Academy of Management Annals,* 5: 1–52.
Beer, M. and Walton, A.E. (1987). Organization change and development. *Annual Review of Psychology,* 38: 339–367.
Binnewies, C., Ohly, S. and Sonnentag, S. (2007). Taking personal initiative and communicating about ideas. *European Journal of Work and Organizational Psychology,* 16(4): 432–455.
Birkinshaw, J., Healey, M.P., Suddaby, R. and Weber, K. (2014). Debating the future of management research. *Journal of Management Studies,* 51(1): 38–55.
Björk, J. and Magnusson, M. (2009). Where do good innovation ideas come from? Exploring the influence of network connectivity on innovation idea quality. *Journal of Product Innovation Management,* 26(6): 662–670.
Blau, P.M. and Schoenherr, R.A. (1971). *The Structure of Organizations.* New York, NY: Basic Books.
Blau, P.M. and Scott, W. (1962). *Formal Organizations.* San Francisco, CA: Chandler.
Blindenbach-Driessen, F.P. and van den Ende, J.C.M. (2010). Innovation management practices compared: The example of product-based firms. *Journal of Product Innovation Management,* 27(5): 705–724.
Blindenbach-Driessen, F.P., Van Dalen, J. and van den Ende, J.C.M. (2010). Subjective performance assessment of innovation projects. *Journal of Product Innovation Management,* 27(4): 572–592.
Bommer, M. and Jalajas, D.S. (1999). The threat of organizational downsizing on the innovative propensity of R&D professionals. *R&D Management,* 29(1): 27–34.
Bonner, J.M., Ruekert, R.W. and Walker, O.C. (2002). Upper management control of new product development projects and project performance. *Journal of Product Innovation Management,* 19(3): 233–245.
Borgatti, S.P. (2005). Centrality and network flow. *Social Networks,* 27(1): 55–71.
Borgatti, S.P. and Cross, R. (2003). A relational view of information seeking and learning in social networks. *Management Science,* 49(4): 432–445.
Borgatti, S.P. and Foster, P. (2003). The network paradigm in organizational research: A review and typology. *Journal of Management,* 29(6): 991–1013.
Bouty, I. (2000). Interpersonal and interaction influences on informal resource exchanges between R&D researchers across organizational boundaries. *Academy of Management Journal,* 43(1): 50–65.
Bovasso, G. (1996). A network analysis of social pontagion processes in an organizational intervention. *Human Relations,* 49(11): 1419–1435.

Brass, D.J. (1984). Being in the right place: A structural analysis of individual influence in an organization. *Administrative Science Quarterly*, 29(4): 518–539

Brass, D.J. and Burkhardt, M.E. (1992). Centrality and power in organizations. In N. Nohria and R.G. Eccles (eds). *Networks and Organizations: Structure, Form and Action* (pp. 191–215). Boston, MA: Harvard Business School Press.

Brass, D.J. and Krackhardt, D. (2012). Power, politics, and social networks. In G.R. Ferris and D.C. Treadway (eds). *Politics in organizations: Theory and research considerations*. New York: Routledge, pp. 355–375.

Brockner, J., Grover, S., Reed, T., DeWitt, R. and O'Malley, M. (1987). Survivors' reactions to layoffs: We get by with a little help for our friends. *Administrative Science Quarterly*, 526–541.

Burt, R.S. (1983). Distinguishing relational contents. In R.S. Burt and M.J. Minor (eds). *Applied Network Analysis*. Beverly Hills, CA: Sage, pp. 35–74.

Burt, R.S. (1984). Network items and general social survey. *Social Network*, 6: 293–339.

Burt, R.S. (1992). *Structural Holes: The Social Structure of Competition*. Cambridge, MA: Harvard University Press.

Burt, R.S. (1997). The contingent value of social capital. *Administrative Science Quarterly*, 42: 339–365.

Burt, R.S. (2004). Structural holes and good ideas. *American Journal of Sociology*, 110(2): 349–399.

Burt, R.S., Jannotta, J.E. and Mahoney, J.T. (1998). Personality correlates of structural holes. *Social Networks*, 20: 63–87.

Campbell, K., Marsden, P. and Hurlbert, J. (1986). Social resources and socioeconomic status. *Social Networks*, 8: 97–117.

Carlile, P.R. (2004). Transferring, translating and transforming: An integrative framework for managing knowledge across boundaries. *Organization Science*, 15(5): 555–568.

Carlile, P.R. and Rebentisch, E.S. (2003). Into the black box: The knowledge transformation cycle. *Management Science*, 49: 1180–1195.

Carroll, G. and Teo, A. (1996). On the social networks of managers. *Academy of Management Journal*, 39: 421–440.

Cohen, W.M. and Levinthal, D.A. (1990). Absorptive capacity: A new perspective on learning and innovation. *Administrative Science Quarterly*, 35: 128–152.

Coleman, J.S. (1988). Social capital in the creation of human capital. *American Journal of Sociology*, 94(1): 95-120.

Coleman, J.S. (1990). *Foundations of Social Theory*. Cambridge, MA: Harvard University Press.

Contractor, N.S., Monge, P.R. and Leonardi, P.M. (2011). Multidimensional networks and the dynamics of sociomateriality: Bringing technology inside the network, *International Journal of Communication* 5: 682–720

Cook, K.S. (1977). Exchange and power in networks of interorganizational relations. *Sociological Quarterly*, 18: 62–82.

Cooper, R.G., Edgett, S.J. and Kleinschmidt, E.J. (1999). New product portfolio management: Practices and performance. *Journal of Product Innovation Management*, 16(4): 333–351.

Cooper, R.G., Edgett, S.J. and Kleinschmidt, E.J. (2004). Benchmarking best NPD practices-I. *Research Technology Management,* 47: 31–43.

Cooper, R.G. and Kleinschmidt, E.J. (1986). An investigation into the new product process: Steps, deficiencies and impact. *Journal of Product Innovation Management,* 3: 71–85.

Cooper, R.G. and Kleinschmidt, E.J. (1995). Benchmarking the firm's critical success factors in product development, *Journal of Product Innovation Management* 12: 374–391.

Cross, R., Borgatti, S.P. and Parker, A. (2002). Making invisible work visible: Using social network analysis to support strategic collaboration. *California Management Review,* 44(2): 25–46.

Cross, R. and Cummings, J.N. (2004). Tie and network correlates of individual performance in knowledge intensive work. *Academy of Management Journal,* 47: 928–937.

Cross R. and Parker, A. (2004) *The Hidden Power of Social Networks: Understanding how work really gets done in Organizations,* Boston, MA: Harvard Business School Publishing.

Cross, R., Parker, A., Prusak, L. and Borgatti, S. (2001). Knowing what we know: Supporting knowledge creation and sharing in social networks. *Organizational Dynamics,* 30(2): 100–120.

Cross, R. and Prusak, L. (2002). The people who make organizations stop – or go. *Harvard Business Review,* 80(6): 104–112.

Cummings, J.N. (2004). Work groups, structural diversity, and knowledge sharing in a global organization. *Management Science,* 50(3): 352–364.

Damanpour, F. (1991). Organizational innovation: A meta-analysis of effect of determinants and moderators, *Academy of Management Journal,* 34(3): 555–590.

Darr, E.D., Argote, L. and Epple, D. (1995). The acquisition, transfer and depreciation of knowledge in service organizations: Productivity in franchises. *Management Science,* 41: 1750–1762.

Datta, D.K., Guthrie, J.P., Basuil, D. and Pandey, A. (2010). Causes and effects of employee downsizing: A review and synthesis. *Journal of Management,* 36(1): 281–348.

Davenport, T.H. and L. Prusak (1998). *Working Knowledge: How Organizations Manage What They Know.* Cambridge, MA: Harvard Business School Press.

Davila, T., Epstein, M. and Shelton, R. (2012). *Making Innovation Work: How to Manage It, Measure It, and Profit from It.* FT press.

DeChurch, L.A. and Marks, M.A. (2006). Leadership in multiteam systems. *Journal of Applied Psychology,* 91: 311–329.

De Lange, D., Agneessens, F. and Waege, H. (2004). Asking social network questions: a quality assessment of different measures. *Metodoloski zvezki,* 1(2): 351–378.

Diehl, M. and Stroebe, W. (1987). Productivity loss in brainstorming groups. *Journal of Personality and Social Psychology,* 53: 497–509.

Dillman, D.A. (1978 [2000]). *Mail and Telephone Surveys: The Total Design Method.* New York: Wiley-Interscience Publication.

Ding, W.W., Levin, S.G., Stephan, P. E. and Winkler, A.E. (2010). The impact of information technology on academic scientists' productivity and collaboration patterns. *Management Science,* 56(9): 1439–1461.

Dolfsma, W.A., van der Eijk, R. and Jolink, A. (2009). On a source of social capital: Gift exchange. *Journal of Business Ethics,* 89(3): 315–329.

Doreian, P. and Stokman, F.N. (2005). The dynamics and evolution of social networks. In: P. Doreian and F.N. Stokman (eds) *Evolution of Social Networks.* New York: Gordon and Breach, 1–17.

Dougherty, D. (1992). Interpretive barriers to successful product innovations in large firms. *Organization Science,* 3(2): 179–202.

Dougherty, D. and Bowman, E.H. (1995). The effects of organizational downsizing on product innovation. *California Management Review,* 37(4): 28-44.

Du Gay, P. (1996). *Consumption and Identity at Work.* London: Sage.

Duncan, R.B. (1976). The ambidextrous organization: Designing dual structures for innovation. In R.H. Kilmann, L.R. Pondy and D. Slevin (eds). *The Management of Organization* (pp. 167–188). New York: North-Holland.

Dyer, J., Gregersen, H. and Christensen, C.M. (2011). *The Innovator's DNA: Mastering the Five Skills of Disruptive Innovators.* Boston, MA: Harvard Business School Press.

Eisenhardt, K.M. (1989). Building theories from case study research. *Academy of Management Review,* 14(4): 532–550.

Ensign, P.C. (2009). *Knowledge Sharing Among Scientists.* New York: Palgrave Macmillan.

Fernandez, R.M. and Gould, R.V. (1994). A dilemma of state power: brokerage and influence in the national-health policy domain, *American Journal of Sociology* 99(6): 1455–91.

Ferriani, S., Fonti, F. and Corrado, R. (2013). The social and economic bases of network multiplexity: Exploring the emergence of multiplex ties. *Strategic Organization,* 11(1): 7–34.

Fisher, S.R. and White, M.A. (2000). Downsizing in a learning organization: Are there hidden costs? *Academy of Management Review,* 25(1): 244–251.

Floyd, S.W. and Lane, P.J. (2000). Strategizing throughout the organization: Managing role conflict in strategic renewal. *Academy of Management Review,* 25(1): 154–177.

Folger, R. and Skarlicki, D. (1998) When tough times make tough bosses: Managerial distancing as a function of layoff blame, *The Academy of Management Journal,* 41(1): 79–87.

Foss, N.J. (2007). The emerging knowledge governance approach: Challenges and characteristics. *Organization,* 14: 29–52.

Foss, N.J., Husted, K. and Michailova, S. (2010). Governing knowledge sharing in organizations. *Journal of Management Studies,* 47(3): 455–482.

Foss, N.J., Laursen, K. and Pedersen, T. (2011). Linking customer interaction and innovation: The mediating role of new organizational practices. *Organization Science,* 22(4): 980–999.

Foss, N.J, Minbaeva, D.B., Pedersen, T. and Reinholt, M. (2009). Encouraging knowledge sharing among employees: How job design matters. *Human Resource Management,* 48(6): 871–893.

Freeman, L.C. (1979). Centrality in social networks. *Social Networks,* 1: 215–239.

Friedman, A.F. and Podolny, J. (1993). Differentiation of boundary spanning roles: Labor negotiations and implications for role conflict. *Administrative Science Quarterly,* 37: 28–47.

Frost, P.J. and Egri, C.P. (1991). The political process of innovation. *Organizational Behavior,* 13: 229–295.

Gabarro, J. (1990). The development of working relationships. In J. Galegher, R.E. Kraut and C. Egido (eds). *Intellectual teamwork: Social and technological foundations of cooperative work.* Hillsdale, NJ: Lawrence Erlbaum, pp. 79–110.

Gandolfi, F. and Oster, G. (2010). How does downsizing impact the innovative capability of a firm? A contemporary discussion with conceptual frameworks. *International Journal of Innovation and Learning,* 8(2): 127–148.

Garvey, C. (2002). Steer teams with the right pay. *HR Magazine,* 19–20.

Gibney, R., Zagenczyk, T.J. and Masters, M.F. (2009). The negative aspects of social exchange: An introduction to perceived organizational obstruction. *Group and Organization Management,* 34(6): 665–697.

Girotra, K., Terwiesch, C. and Ulrich, K. (2010). Idea generation and the quality of the best idea, *Management Science* 56(4): 591–605.

Gittell, J.H., Cameron, K., Lim, S. and Rivas, V. (2006). Relationships, layoffs, and organizational resilience. *The Journal of Applied Behavioral Science,* 42(3): 300–329.

Giuliani, E. and Bell, M. (2005). The micro-determinants of meso-level learning and innovation: evidence from a Chilean wine cluster, *Research Policy,* 34(1): 47–68.

Goodwin, V.L., Bowler, W.M. and Whittington, J.L. (2008). A social network perspective on LMX relationships. *Journal of Management,* 35: 964–980.

Gould, R.V. and Fernandez, R. (1989). Structures of mediation: A formal approach to brokerage in transaction networks. *Sociological Methodology,* 19: 89–126.

Granovetter, M.S. (1973). The strength of weak ties. *American Journal of Sociology,* 78(6): 1360–1380.

Granovetter, M. (1982). The strength of weak ties: a network theory revisited. In: Marsden, P. and Lin, N. (eds) *Social structure and network analysis.* Beverley Hills, CA: Sage.

Granovetter, M. (1995). *Getting a Job: A Study of Contacts and Careers.* Chicago, IL: University of Chicago Press (2nd ed.).

Grant, R.M. (1996). Toward a knowledge-based theory of the firm. *Strategic Management Journal,* 17(S2): 109–122.

Griffin, A. (1997). PDMA research on new product development practices. *Journal of Product Innovation Management,* 14: 429–458.

Grosser, T., Lopez-Kidwell, V., Labianca, G. and Ellwardt, L. (2012). Hearing it through the grapevine: Positive and negative workplace gossip. *Organizational Dynamics,* 41: 42–56.

Gulati, R. (1995). Social structure and alliance formation patterns: A longitudinal analysis. *Administrative Science Quarterly,* 40: 619–652.

Gulati, R. and Puranam, P. (2009). Renewal through reorganization: The value of inconsistencies between formal and informal organization. *Organization Science,* 20(2): 422–440.

Gulati, R., Sytch, M. and Tatarynowicz, A. (2012). The rise and fall of small worlds: Exploring the dynamics of social structure. *Organization Science,* 23(2): 449–471.

Gupta, A.K. and Govindarajan, V. (2000). Knowledge flows within multinational corporations. *Strategic Management Journal,* 21: 473–496.

Haas, M.R. (2010). The double-edged swords of autonomy and external knowledge:

Team effectiveness in a multinational organization. *Academy of Management Journal,* 53: 989–1008.

Haas, M.R. and Hansen, M.T. (2005). When using knowledge can hurt performance: Value of organizational capabilities in a management consulting company. *Strategic Management Journal,* 26(1): 1–24.

Han, S.K. (1996). Structuring relations in on-the-job networks. *Social Networks,* 18(1): 47–67.

Hansen, M.T. (1999). The search-transfer problem: The role of weak ties in sharing knowledge across organization subunits. *Administrative Science Quarterly,* 44: 82–111.

Hansen, M.T. (2002). Knowledge networks: Explaining effective knowledge sharing in multiunit companies. *Organization Science,* 13(3): 232–248.

Hansen, M.T. and Birkinshaw J. (2007) The innovation value chain. *Harvard Business Review,* 85(6): 121–130.

Hansen, M.T. and Lovas, B. (2004). How do multinational companies leverage technological competencies? Moving from single to interdependent explanations. *Strategic Management Journal,* 25: 801–822.

Hansen, M.T., Mors, M.L. and Lovas B. (2005). Knowledge sharing in organizations: multiple networks, multiple phases. *Academy of Management Journal,* 48(5): 776–793.

Hansen, M.T., Nohria, N. and Thomas T. (1999). What's your strategy for managing knowledge? *Harvard Business Review,* 77(2): 106–116.

Hansen, M.T., Podolny, J.M. and Pfeffer, J. (2001). So many ties, so little time: A task contingency perspective on corporate social capital. *Research in the Sociology of Organizations,* 18: 21–57.

Hargadon, A.B. (2002). Brokering knowledge: Linking learning and innovation. *Research in Organizational Behavior,* 24: 41–85.

Hargadon, A.B. and Bechky, B.A. (2006). When collections of creatives become creative collectives: A field study of problem solving at work, *Organization Science,* 17(4): 484–500.

Harrisson, D. and Laberge, M. (2002). Innovation, identities and resistance: The social construction of an innovation network. *Journal of Management Studies,* 39(4): 497–521.

Hartman, R.L. and Johnson, J.D. (1979). Social contagion and multiplexity. Communication networks as predictors of commitment and role ambiguity. *Human Communication Research,* 15(4): 523–548.

Henry, R.A. (1995). Improving group judgment accuracy: Information sharing and determining the best member. *Organizational Behavior and Human Decision Processes,* 62(2): 190–197.

Hershock, R.J., Cowman, C.D. and Peters, D. (1994). Action teams that work. *Journal of Product Innovation Management,* 11(2): 95–104.

Hesse-Biber, S.N. and Leavy, P. (2010). *The Practice of Qualitative Research.* Sage.

Hoang, H. and Antoncic, A. (2003). Network-based research in entrepreneurship: A critical review, *Journal of Business Venturing,* 18(2): 165–187.

Ibarra, H. (1993). Network centrality, power and innovation involvement: Determinants of technical and administrative roles. *Academy of Management Journal,* 36(3): 471–501.

Ibarra, H. (1995). Race, opportunity, and diversity of social circles in managerial networks. *Academy of Management Journal,* 38: 673–703.

Ibarra, H. and Andrews, S.B. (1993). Power, social influence, and sense making: Effects of network centrality and proximity on employee perceptions, *Administrative Science Quarterly,* 38: 277–303.

Kahn, R.L., Wolfe, D.M., Quinn, R.P., Snoek, J.D. and Rosenthal, R.A. (1964). *Organizational stress: Studies in role conflict and ambiguity.* New York: Wiley.

Kalish, Y. and Robins, G.L. (2006). Psychological predispositions and network structure: The relationship between individual predispositions, structural holes and network closure. *Social Networks,* 28: 56–84.

Katz, D. and Kahn, R.L. (1978). *The Social Psychology of Organizations.* New York: Wiley.

Kelley, D. and Lee, H. (2010). Managing innovation champions: The impact of project characteristics on the direct manager role. *Journal of Product Innovation Management,* 27(7): 1007–1019.

Kijkuit, B. and van den Ende, J. (2007). The organizational life of an idea: Integrating social network, creativity and decision-making perspectives. *Journal of Management Studies,* 44(6): 860–882.

Kijkuit, B. and van den Ende, J. (2010). With a little help from our colleagues: A longitudinal study of social networks for innovation. *Organization Studies,* 31(4): 451–479.

Kilduff, M. and Brass, D.J. (2001). The social network of high and low self-monitors: Implications for workplace performance. *Administrative Science Quarterly,* 46(1): 121–146.

Kleinbaum, A.M. (2012). Organizational misfits and the origins of brokerage in intrafirm networks. *Administrative Science Quarterly,* 57(3): 407–452.

Kleinbaum, A.M. and Tushman, M.L. (2007). Building bridges: The social structure of interdependent innovation. *Strategic Entrepreneurship Journal,* 1(1–2): 103–122.

Kohn, N.W., Paulus, P.B. and Choi, Y. (2011). Building on the ideas of others: An examination of the idea combination process. *Journal of Experimental Social Psychology,* 47(3): 554–561.

Krackhardt, D. and Hanson, J. (1993). Informal networks: The company behind the chart. *Harvard Business Review,* 71: 104–11.

Kratzer, J., Leenders, R.Th.A.J. and van Engelen, J.M.L. (2010). The social network among engineering design teams and their creativity. *International Journal of Project Management,* 28: 428–436.

Lane, P.J. and Lubatkin, M. (1998). Relative absorptive capacity and interorganizational learning. *Strategic Management Journal,* 19: 461–477.

Lazega, E. and Pattison, P.E. (1999). Multiplexity, generalized exchange and cooperation in organizations: A case study. *Social Networks,* 21(1): 67–90.

Lee, J. (2010). Heterogeneity, brokerage, and innovative performance: Endogenous formation of collaborative inventor networks. *Organization Science,* 21: 804–822.

Lee, J.-Y., Bachrach, D. G. and Lewis, K. (2014). Social Network Ties, Transactive Memory, and Performance in Groups. *Organization Science,* 25(3): 951–967. doi:10.1287/orsc.2013.0884

Leenders, R.Th.A.J. van Engelen, J.M.L. and Kratzer, J. (2003). Virtuality, communication, and new product team creativity: A social network perspective. *Journal of Engineering and Technology Management*, 20: 69–92.

Leenders, R.Th. A.J. van Engelen, J.M.L. and Kratzer, J. (2007a). Systematic design methods and the creative performance of new product teams: Do they contradict or complement each other? *Journal of Product Innovation Management*, 24: 166–179.

Leenders, R.Th.A.J., Kratzer, J. and van Engelen, J.M.L. (2007b). Innovation team networks: The centrality of innovativeness and efficiency. *International Journal of Networking and Virtual Organizations*, 4: 459–478.

Levin, D.Z. and Cross, R. (2004). The strength of weak ties you can trust: The mediating role of trust in effective knowledge transfer. *Management Science*, 50(11): 1477–1490.

Levine, J.M., Moreland, R.L. and Choi, H.S. (2001). Group socialization and newcomer innovation. In M. Hogg and S. Tindale, (eds). *Blackwell Handbook in Social Psychology* (Vol. 3, Group Processes. Oxford: Blackwell Publishers Limited, pp. 86–106.

Levine, S.S. and Prietula, M. (2011). How knowledge transfer impacts performance: A multi-level model of benefits and liabilities. *Organization Science*, 23: 1748–1766.

Li, Q., Maggitti, P.G., Smith, K.G., Tesluk, P.E. and Katila, R. (2013). Top management attention to innovation: The role of search selection and intensity in new product introductions. *Academy of Management Journal*, 56(3): 893–916.

Lincoln, J.R. and Miller, J. (1979). Work and friendship ties in organizations: A comparative analysis of relational networks. *Administrative Science Quarterly*, 24: 181–199.

Littler, C.R. (2000). Comparing the downsizing experiences of three countries: A restructuring cycle? In R.J. Burke and C.L. Cooper (eds). *The Organization in Crisis*. Malden, MA: Blackwell Publishers, pp. 58–77.

Logan, M.S. and Ganster, D.C. (2007). The effects of empowerment on attitudes and performance: The role of social support and empowerment beliefs. *Journal of Management Studies*, 44(8): 1523–1550.

Madhaven, R. and Grover, R. (1998). From embedded knowledge to embodied knowledge: New product development as knowledge. *Management Journal of Marketing*, 62: 1–12.

Malik, K. (2002). Aiding the technology manager: A conceptual model for intra-firm technology transfer. *Technovation*, 22: 427–436.

March, J.G. (1991). Exploration and exploitation in organizational learning. *Organization Science*, 2(1): 71–87.

Markham, S.K. (1998). A longitudinal examination of how champions influence others to support their projects. *Journal of Product Innovation Management*, 15: 490–504.

Marrone, J.A. (2010). Team boundary spanning: A multi-level review of past research and proposals for the future. *Journal of Management*, 36(4): 911–940.

Marrone, J.A., Tesluk, P.E. and Carson, J.B. (2007). A multilevel investigation of antecedents and consequences of team member boundary-spanning behavior. *Academy of Management Journal*, 50(6): 1423–1439.

Marsden, P.V. (1981). Introducing influence processes into systems of collective decisions. *American Journal of Sociology,* 86: 1203–1235.
Marsden, P.V. (1982). Brokerage behavior in restricted exchange networks. *Social Structure and Network Analysis,* 7(4): 341–410.
Marsden, P.V. (1990). Network data and measurement. *Annual Review of Sociology,* 16: 435–463.
Marsden, P.V. (2002). Egocentric and sociocentric measures of network centrality. *Social Networks,* 24(4): 407–422.
Matusik, S.F. and Heeley, M.B. (2005). Absorptive capacity in the software industry: Identifying dimensions that affect knowledge and knowledge creation activities. *Journal of Management,* 31(4): 549–572.
Maute, M.F. and Locander, W.B. (1994). Innovation as a socio-political process: An empirical analysis of influence behavior among new product managers. *Journal of Business Research,* 30: 161–174.
McEvily, B., Perrone, V. and Zaheer, A. (2003). Trust as an organizing principle. *Organization Science,* 14: 91–103.
McLaughlin, S., Paton, R.A. and Macbeth, D.K. (2008). Barrier impact on organizational learning within complex organizations. *Journal of Knowledge Management,* 12(2): 107–123.
Mehra, A., Kilduff, M. and Brass, D.J. (2001). The social networks of high and low self-monitors: Implications for workplace performance. *Administrative Science Quarterly,* 46: 121–146.
Merton, R.K. (1968). *Social Theory and Social Structure.* New York: Free Press.
Minor, M.J. (1983). New directions in multiplexity analysis. In: R.S. Burt and M.J. Minor (eds). *Applied Network Analysis.* Beverly Hills, CA: Sage, pp. 223–244.
Mintzberg, H. (1980). Structure in 5's: A synthesis of the research on organization design. *Management Science,* 26: 322–341.
Mishra, K.E., Spreitzer, G.M. and Mishra, A.K. (1998). Preserving employee morale during downsizing, *MIT Sloan Management Review,* 39(2): 83–95.
Mizruchi, M.S. and Brewster Stearns, L. (2001). Getting deals done: The use of social networks in bank decision-making. *American Sociological Review,* 66: 647–671.
Moenaert, R.K., Caeldries, F., Lievens, A. and Wauters, E. (2000). Communication flows in international product innovation teams. *Journal of Product Innovation Management,* 17: 360–377.
Mom, T.J.M., van den Bosch, F.A.J. and Volberda, H.W. (2009). Understanding variation in managers' ambidexterity: Investigating direct and interaction effects of formal structural and personal coordination mechanisms. *Organization Science,* 20(4): 812–828.
Moorman, C. and Miner, A.S. (1998). Organizational improvisation and organizational memory. *Academy of Management Review,* 23: 698–723.
Moran, P. (2005). Structural vs. Relational Embeddedness: Social capital and managerial performance. *Strategic Management Journal,* 26: 1129–1151.
Moreno, J.L., (1961). Role concept, a bridge between psychiatry and sociology. *American Journal of Psychiatry,* 118: 518–523.
Mors, M.L. (2010). Innovation in a Global Consulting Firm: When the problem is too much diversity. *Strategic Management Journal,* 31(8): 841–872.

Nahapiet, J. and Ghoshal, S. (1998). Social capital, intellectual capital, and the organizational advantage. *Academy of Management Review*, 23(2): 242–266.

Newman, M.E.J. (2003). Ego-centered networks and the ripple effect. *Social Networks*, 25: 83–95.

Nixon, R.D., Michael, A.H., Ho-Uk, L. and Eui, J. (2004). Market reactions to announcements of corporate downsizing actions and implementation strategies. *Strategic Management Journal*, 25(11): 1121–1129.

Nonaka, I. (1994). A dynamic theory of organizational knowledge creation. *Organization Science*, 5(1): 14–37.

Nonaka, I. and Teece, D. (2001). *Managing Industrial Knowledge: Creation, Transfer and Utilization*. Sage.

Nooteboom, B. (2002). *Trust: Forms, Foundations, Functions, Failures and Figures*. Cheltenham UK: Edward Elgar.

Obstfeld, D. (2005). Social networks, the tertius lungens orientation, and involvement in innovation. *Administrative Science Quarterly*, 50: 100–130.

Ocasio, W. (1995). The enactment of economic adversity: A reconciliation of theories of failure-induced change and threat-rigidity. In L.L. Cummings and B.M. Staw (eds). *Research in Organizational Behaviour*. Greenwich, CT: JAI Press, 17: 287–331.

Ocasio, W. (1997). Towards an attention-based view of the firm. *Strategic Management Journal*, 18: 187–206.

Oh, H. and Kilduff, M. (2008). The ripple effect of personality on social structure: Self-monitoring origins of network brokerage. *Journal of Applied Psychology*, 93: 1155–1164.

Ohly, S., Kase, R. and Škerlavaj, M. (2010). Networks for generating and validating ideas. *Innovation: Management, Policy and Practice*, 12(1): 50–60.

Okhuysen, G.A. (2001). Structuring change: Familiarity and formal interventions in problem-solving groups. *Academy of Management Journal*, 44: 794–808.

Okhuysen, G.A. and Eisenhardt, K.M. (2002). Integrating knowledge in groups: How formal interventions enable flexibility. *Organization Science*, 13(4): 370–386.

Parise, S., Cross, R. and Davenport, T.H. (2006). Strategies for preventing a knowledge-loss crisis. *MIT Sloan Management Review*, 47(4): 31–38.

Paulus, P. (2000). Groups, teams, and creativity: The creative potential of idea-generating groups. *Applied psychology: An International review*, 49, 237–262.

Perry-Smith, J.E. and Shalley, C.E. (2003). The social side of creativity: A static and dynamic social network perspective. *Academy of Management Review*, 28(1): 89–107.

Porter, L.W. and Lawler, E.E. (1968). *Managerial Attitudes and Performance*. Homewood, IL: Richard D. Irwin.

Porter, L.W., Allen, R.W. and Angle, H.L. (1981). The politics of upward influence in organizations. In L. Cummings and B. Staw (eds). *Research in Organizational Behaviour*. Greenwich, England: Elsevier Ltd, pp. 109–149.

Powell, W.W., K. Koput and Smith-Doerr L. (1996). Interorganizational collaboration and the locus of innovation: Networks of learning in biotechnology, *Administrative Science Quarterly* 41(1): 116–45.

Provan, K.G., Fish, A. and Sydow, J. (2007). Interorganizational networks at the network level: A review of the empirical literature on whole networks. *Journal of Management*, 33(3): 1–65.

Reagans, R. and McEvily, B. (2003). Network structure and knowledge transfer: The effects of cohesion and range. *Administrative Science Quarterly,* 48(2): 240–267.

Rizova, P.S. (2007). *The Double Helix of Formal and Informal Structures in an R&D Laboratory.* Stanford, CA: Stanford UP.

Robertson, P.J., Roberts, D.R. and Porras, J.I. (1993). An evaluation of a model of planned organizational change. In R.W. Woodman and W.A. Pasmore (eds). *Research in Organizational Change and Development.* Greenwich, CT: JAI Press, 7: 1–39.

Robertson, M. and Swan, J. (2003). Control – what control... Culture and ambiguity within a knowledge intensive firm. *Journal of Management Studies,* 40(4): 831–858.

Robins, G.L. and Pattison, P. (2006). *Multiple Networks in Organisations.* Report to DSTO.

Rodan, S. (2010). Structural holes and managerial performance: Identifying the underlying mechanisms. *Social Networks,* 32(3): 168–179.

Rogers, E.M. and Kincaid, D.L. (1981). *Communication Networks: Toward a New Paradigm for Research.* New York: Free Press.

Roth, K. and Kostova, T. (2003). The use of the multinational corporation as a research context. *Journal of Management,* 9(6): 883–890.

Saxenian, A. (1994) *Regional Advantage: Culture and Competition in Silicon Valley and Route 128.* Cambridge, MA: Harvard University Press.

Schilit, W.K. (1986). An examination of individual differences as moderators of upward influence activity in strategic decisions. *Human Relations,* 39(10): 933–953.

Schoonhoven, C.B. and Jellinek, M. (1990). Dynamic tension in innovative, high technology firms: Managing rapid technology change through organization structure. In M.A. Von Glinow and S.A. Mohrman (eds). *Managing Complexity in High Technology Organizations.* New York: Oxford UP, pp. 90–118.

Schulz, M. (2003). Pathways of relevance: Exploring inflows of knowledge into subunits of multinational corporations. *Organization Science,* 14(4): 440–459.

Sethia, N.K. (1995). The role of collaboration in creativity. In C.M. Ford and D.A. Goia (eds). *Creative Action in Organization.* Thousand Oaks, CA: Sage, pp. 100–105.

Shah, P.P. (2000). Network destruction: The structural implications of downsizing. *Academy of Management Journal,* 43(1): 101–112.

Shim, D. and Lee, M. (2001). Upward influence styles of R&D project leaders. *IEEE Transactions on Engineering Management,* 48(4): 394–413.

Sias, P.M. and Cahill, D.J. (1998). From coworkers to friends: The development of peer friendships in the workplace. *Western Journal of Communication,* 62(3): 273–299.

Simmel, G. (1950). *The Sociology of Georg Simmel.* New York: Free Press, 1st Free Press paperback edition.

Simon, H.A. (1976). *Administrative Behavior.* New York: Free Press.

Singh, J. (2005). Collaborative networks as determinants of knowledge diffusion patterns. *Management Science,* 51(5): 756–770.

Singh, J. and Fleming, L. (2010). Lone inventors as sources of technological breakthroughs: Myth or reality? *Management Science,* 56: 41–56

Sivadas, E. and Dwyer, F.R. (2000). An examination of organizational factors influencing new product success in internal and alliance-based processes. *Journal of Marketing*, 64: 31–49.

Smith, J.M., Halgin, D.S., Kidwell-Lopez, V., Labianca, G., Brass, D.J. and Borgatti, S.P. (2014). Power in politically charged networks. *Social Networks*, 36: 162–176.

Smith-Doerr, L., Manev, I. and Rizova, P. (2004). The meaning of success: Network position and the social construction of project success in an R&D lab. *Journal of Engineering and Technology Management*, 21(1–2): 51–81.

Smith-Doerr, L. and Powell, W.W. (2005). Networks and economic life. In N.J. Smelser and R. Swedberg, (eds). *The Handbook of Economic Sociology*. Princeton, NJ: Russell Sage Foundation/Princeton UP, pp. 379–402.

Soda, G. and Zaheer, A. (2012). A network perspective on organizational architecture: performance effects of the interplay of formal and informal organization. *Strategic Management Journal* 33(6): 751–771,

Stevenson, W.B. and Gilly, M.C. (1991). Information processing and problem solving: The migration of problems through formal positions and networks of ties. *Academy of Management Journal*, 34, 918–928.

Subramaniam, M. and Youndt, M.A. (2005). The influence of intellectual capital on the nature of innovative capabilities. *Academy of Management Journal*, 48(3): 450–464.

Sudman, S. and Bradburn, N. (1982) *Asking Questions: a Practical Guide to Questionnaire Design*. USA: Jossey-Bass.

Szulanski, G. (1996). Exploring internal stickiness: Impediments to the transfer of best practice within the firm, *Strategic Management Journal*, 17, 27–43.

Szulanski, G. (2003). *Sticky Knowledge: Barriers to Knowing in the Firm*. London: Sage.

Szulanski, G., Cappetta, R., Jensen, R.J. (2004). When and how trustworthiness matters: Knowledge transfer and the moderating effect of causal ambiguity. *Organization Science*, 15(5): 600–613.

Taube, V. (2003). Measuring the social capital of brokerage roles. *Connections*, 25(2): 1–25.

Teece, D., Pisano, G. and Shuen, A. (1997). Dynamic capabilities and strategic management. *Strategic Management Journal*, 18(7): 509–533.

Teigland, R. and Wasko, M. (2009). Knowledge transfer in MNCs: Examining how intrinsic motivations and knowledge sourcing impact individual centrality and performance. *Journal of International Management*, 15(1): 15–31.

Thornton, P. and Flynn, K. (2003). Entrepreneurship, networks and geographies. In Z.J. Acs and D.B. Audretsch (eds). *Handbook of Entrepreneurship Research*. Boston, MA: Kluwer Academic Publishers, pp. 401–433.

Tortoriello, M. and Krackhardt, D. (2010). Activating cross boundary knowledge: The role of Simmilian ties in the generation of innovation. *Academy of Management Journal*, 53(1): 167–181.

Tourangeau, R., Rips, L.J. and Rasinski, K.A. (2000). *The Psychology of Survey Response*. Cambridge: Cambridge University Press.

Tourangeau, R. and Smith, T.W. (1996). Asking sensitive questions: The impact of data collection mode, question format, and question context, *Public Opinion Quarterly*, 60(2): 275–304.

Tourish, D. and Pinnington, A.H. (2002). Transformational leadership, corporate cultism and the spirituality paradigm: An unholy trinity in the workplace? *Human Relations,* 55(2): 147–172.

Tsai, W. (2000). Social capital, strategic relatedness, and the formation of intra-organizational strategic linkages. *Strategic Management Journal,* 21: 925–939.

Tsai, W. (2001). Knowledge transfer in intraorganizational networks: Effects of network position and absorptive capacity on business unit innovation and performance. *Academy of Management Journal,* 44: 996–1004.

Tsai, W. (2002). Social structure of coopetition within a multiunit organization: Coordination, competition, and intra-organizational knowledge sharing. *Organization Science,* 13: 179–190.

Tushman, M. (1977). Special boundary roles in the innovation process. *Administrative Science Quarterly,* 22(4): 587–605.

Tushman, M. (1979). Impacts of perceived environmental variability on patterns of work related communication. *Academy of Management Journal,* 22(1): 482–500.

Tushman, M.L. and O'Reilly, C.A. (2013). *Winning Through Innovation: A Practical Guide to Leading Organizational Change and Renewal.* Cambridge, MA: Harvard Business Press.

Tyre, M.J. and Orlikowski, W.J. (1994). Windows of opportunity: Temporal patterns of technological adaptation in organizations. *Organization Science,* 5(1): 98–118.

Van der Panne, G. (2004). Agglomeration externalities: Marshall versus Jacobs, *Journal of Evolutionary Economics,* 14: 593–604.

Varella, P., Javidan, M. and Waldman, D.A. (2012). A model of instrumental networks: The roles of socialized charismatic leadership and group behavior. *Organization Science,* 23(2): 582–595.

Von Hippel, E. (1994). Sticky information and the locus of problem solving: Implications for innovation. *Management science,* 40(4): 429–439.

Wagner, K., Foo, E., Zablit, H. and Taylor, A. (2014). *The Most Innovative Companies 2014: Breaking Through is Hard to Do.* Boston Consulting Group Report, October, 2014.

Wasserman, S. and Faust, K. (1994). *Social Network Analysis: Methods and Applications.* New York: Cambridge UP.

Weick, K.E. and Robert, K. (1994). Collective mind in organizations: Heedful interrelating on flight decks. *Administrative Science Quarterly,* 38: 357–381.

Weissenberger-Eibl, M.A. and Teufel, B. (2011). Organizational politics in new product development project selection. *European Journal of Innovation Management,* 14(1): 51–73.

Whelan, E., Parise, S., der Valk, J. and Aalbers, H.L. (2011). Creating employee networks that deliver open innovation. *MIT Sloan Management Review,* 53(1): 37–44.

Whelan, E., Teigland, R., Donnellan, B. and Golden, W. (2010). How internet technologies impact information flows in R&D: Reconsidering the technological gatekeeper. *R&D Management,* 40(4): 400–413.

Winter, S.G. and Szulanski, G. (2001). Replication as Strategy. *Organization Science,* 12: 730–743.

Woodman, R.W., Sawyer, J.E. and Griffin, R.W. (1993). Toward a theory of organizational creativity. *Academy of Management Review,* 18(2): 293–321.

Yin, R.K. (1994). *Case Study Research: Design and Methods.* Thousand Oaks, CA: Sage.

Zagenczyk, T.J., Givney, R., Few, W.T. and Scott, K.L. (2011). Psychological contract and organizational identification: The mediating effect of perceived organizational support, *Journal of Labor Research,* 32: 254–281.

Zellmer-Bruhn, M.E. (2003). Interruptive events and team knowledge acquisition. *Management Science,* 49(4): 514–528.

INDEX

Adler, P. S. 36, 67, 103; formalization of communications 148
advice networks: overview 132
Agarwal, R. 115
agency arguments 38
Agneessens, F. 65
Ahuja, G. 36, 38, 39, 102
Aiken, M. 66
Albrecht, T. L.: transfer of ideas in informal networks 69
Alcacer, J. 36
Allen, T. J. 43, 65, 66, 83, 105; specialization at innovative team level 88; technology gatekeeper 40
Ancona, D. 38, 78, 79, 80, 83, 87, 108
asymmetrical information 103

Bartlett, C. A. 1
Bell, M. 12
betweenness benefits 120, 121
betweenness centrality 13, 14, 30, 121
Blau, P. M.: informal relations within organizations 68
Boeing 118
in-Bonachich power 30
Bonacich centrality 14

bonuses 58
Borgatti, S. P. 64, 130, 133, 152
Borys, B. 36, 67, 103; formalization of communications 148
Boston Consulting Group Innovation Survey 4
boundaries: bridging 55–8; crossing 35, 83
boundary positions 39
boundary spanners 40
Bowman, E. H.: downsizing 118
Brass, D. J. 8, 13, 16, 64, 65, 86, 103, 151
brokerage: agency arguments 38; connecting, exploring, sponsoring 46–8; empirical evidence 43–6; executive 47; external orientation 48; hierarchical 48; identifying employee profiles 36–7; internal orientation 48; knowledge transfer 150; organizational gain 36; profiles 48; restricted to senior staff 49; SNL (Siemens NL) 153; Triad based 41
brokers 11, 32; connecting 47; executive 43, 47; exploring across unit boundaries 38; externally oriented 42; internally oriented 42–3,

172 Index

118–19, 122; knowledge creation 36; knowledge leaders 46; strategic orientation 37; types of 35; unlocking innovation potential 36
BU 1 (Transport and Distribution) 54, 55, 59, 60, 61
BU 2 (Information and Communications) 54, 55, 59, 60
BU 3 (Traffic and Safety) 54, 55, 59
BU 4 (Mobile Communications) 54
Burt, Ronald: brokers 11; managing innovation networks 6
business units: boundaries 36; knowledge transfer 37; Siemens 52; SNL (Siemens NL) 43, 54

Caldwell, D. 38, 78, 79, 80, 83, 87, 108
case studies: data collection 129–30; Greenwood 70, 71–3; Redrock 88–95, 110–13
centrality: assessing knowledge leaders 28; betweenness 13, 14, 30, 121; in-Bonachich power 30; Bonacich 14, 30; in-closeness 30; closeness 13, 14; in-degree 13, 14, 30; degree 152; high-profile persons 150; identification of highly central persons 15; influence of individuals 151; measures of 151; out-Bonachich power 30; out-closeness 30; out-degree 13, 14, 30; sensitivity to network size 15; significance of 13; snapshots of networks 15
Chief Information Officers (CIO's) 143 in-closeness centrality 30
closeness centrality 13, 14, 152–3
clustering 12–13; of expertise 37; SNL (Siemens NL) 59–61
coalition building 16–17
Cohen, S. I. 43, 65, 66
Coleman, J. S.: redundancy in networks 12
combining networks *see* rich ties
commitment: need identification 127–8
communication: BU 2 (Information and Communications) 54, 55; cost of 22, 46; crossing team boundaries 83; directors 143; flows in SNL (Siemens NL) 56, 58; formalization 148; lower costs through intervention 106; networks 58–62; patterns 144; products by SNL 54; profiles 33, 46; reciprocal 61–2, 148–9
communication roles: Consultants 40, 41, 46; Coordinators 40, 41, 42, 46; Gatekeepers 40, 41, 46; Liaison 40, 41, 46; Representatives 40, 41, 46; *see also* network roles
community determination 25
competitive advantage: network of contacts 5; value of knowledge 4
concentrated horizontal and vertical cross-ties 95–6
conferences 47
connections: access to information 5; of employees 22; formal 64; IBM Connections 140; identifying team member to establish 123; informal/third-party 26, 64; of newcomers 27; potential per network 6; *see also* contacts
Consultants 40, 41, 46
contacts: competitive advantage 5; cross-unit 108; diverse 108–9; increasing 107–8; informal 26, 121–2; *see also* connections
cooperation: between business units 43; between divisions 55; between employees 111; *see also* Siemens; SNL (Siemens NL)
Coordinators 40, 41, 42, 46
corporate restructuring 115
corporate themes 54
creativity: benefits of horizontal cross-ties 93; collective 83–4; effective knowledge transfer 4; enhanced by increasing contacts 108; hampering 67; hierarchical brokerage 48; stimulating 83; strong competitive position 53

cross-hierarchy ties 80–2
crossing boundaries 35
cross-ties: concentrated horizontal and vertical 95–6; cross-hierarchy 80–2; cross-unit 80–2; differentiated benefits 92–5; formally and informally 81; horizontal 80, 81–2, 83–4, 88–90, 92–3, 95–6; maintained by limited number of team members 88; network efficiency 87–8; project performance 80; Redrock case study 88–95; relevance of horizontal ties 83–4; relevance of vertical ties 84–7; trendsetting 86; vertical 80, 82, 84–7, 91–3, 94, 95–6; *see also* project teams
cross-unit ties 80–2
customer knowledge networks: overview 132

Damanpour, F. 73; formal organizations 67
data collection: case studies 129–30; data scoping 128–9; designing the process 129–33; ego-centric sampling 130, 131, 133; emergent network 134; frequency of measurement 129; geographical distribution 129; interaction channel 129; isolates 131; level of detail 129; name generators 131–2, 133–4, 136; network boundary 129; network size 129; organizational network questions 133–4; other data 134; respondents' willingness to reply 133; snowball procedure 131, 133; socio-centric sampling 130–1
data network 28
data scoping 128–9
degree centrality 14, 152
diagnosing organizations: approaches 23–4; assessing the climate for innovation 24–7; best structures 27–30; ideation implementation 31–3; knowledge exchange in networks 22–3; *see also*

Organizational Network Analysis (ONA)
directed ties 9
diverse contacts 108–9
diversity 120–1, 144
DNA: innovation 118
domains, knowledge 154
Dougherty, D.: downsizing 118
downsizing: betweenness benefits 120, 121; contribution of well-connected employees 120; corporate restructuring 115; identifying innovation guardians 119–20; impact on innovation 115, 116, 117; impact on product innovation 118; improving efficiency 115; leaders involvement with innovation 121; *see also* reorganization

ego-centric sampling 130, 131, 133
ego level 28, 30
Eisenhardt, K. M. 106, 109, 110; experimental study 105; simple interventions 104
emergent network 134
employees: acquisition of power 8; boundary spanners 40; centrality 13; concept of centrality 15; connections 22; contribution to innovation after downsizing 120; criteria for successful innovation 102; increasing number of contacts 107–8; internally and externally oriented 37–9, 40; knowledge transfer 65; multiplex ties 64; network tactics 16; newcomers 27; particular communication profiles 33; perspectives on knowledge sharing at SNL 56–7; project teams 78–9, 80; reduction of autonomy in formal networks 67; relating to each other in different ways 64; reliance on established relations 87; routines 106; social embeddedness 31; well-connected 120; window of opportunity 32; *see also*

communication roles; nodes
Enterprise Social Software (ESS) 140
entrepreneurs: and networking 5
ethics: research 140–1
exchange: innovation networks 21–2, 25
executive brokers 43, 47
executive brokership 32
external orientation 37–9, 41, 42–3, 44, 118–19; brokerage 48; brokers 42

formal connections 64
formal cross-ties 81
formal networks: formal structures 67–8; formal ties 70; influencing 148; intervention of management 64; organizationally mandated relations 65; overview 132; reduction in autonomy of individuals 67; rich ties 66–8; Siemens 44; SNL (Siemens NL) 58, 59, 147–8; stifling creativity 67
formal organizations 67
formal relations: knowledge transfer 74
formal structures 67–8
formal ties 64, 70, 72
Fortune 1,000 78
Foss, N. J.: formal organizations 67
Freeman, L. C.: basic network times 151–2
funding: management control of 31

Garvey, C. 78
Gatekeepers 40, 41, 46
geographical breakdown 25, 26
Ghoshal, S. 1
Giuliani, E. 12
goals: need identification 127
gossip 16
Granovetter, M. S.: relational embeddedness 11; weak ties 11
grapevine 58, 68, 134
Greenwood: divisional structure 70; formal ties 72, 76; frequency of tie types 72; informal ties 72, 76; innovation network 71; multiplexity as key to exchange of innovative knowledge 73; multiplex ties 72, 76; QAP regression 76; strategic multidisciplinary themes 70; tie types in relation to knowledge transfer 71
guardians 119–20
Gulati, R. 38, 64, 65, 66, 67, 68, 71, 87, 115

Hage, J. 66
Hansen, M. T.: informal relations 69
Hans (SNL change manager) 55–6, 58, 59–62, 149–50
Hargadon, A. B.: knowledge domains 154
Hartman, R. L. 69
Helfat, C. E. 115
Hershock, R. J.: success of project teams 84
Hewlett-Packard 118
hierarchical brokerage 48
high density 12, 22
high redundancy networks 12
horizontal cross-ties 80, 81–2; benefits 92–3; concentrated 95–6; at Redrock 88–90; relevance of 83–4
hubs, knowledge 24

IBM 118
IBM Connections 140
ideas: correlation with employees' absolute number of relations 107–8; developing new 107, 108; evaluating 107; ideation implementation 31–3; interaction and cross-fertilization of 83; managerial intervention 32, 103; political support for 107; resources and funding 31; social embeddedness 31; supportive environment for innovation 32; transferring in informal networks 69; understanding an organization's social infrastructure 31–2
IES (innovation engagement scan): community determination 25;

geographical breakdown 25, 26; ideation generation 31; improving relations between departments 25–6; linchpin 25; objective insights 24–5; outcomes 26, 28; reaction to events 25; subgroups 25
incentive structures 55, 58
in-degree centrality 13, 30
individual ego level 29, 30
individuals *see* employees
influencers 28, 86
informal close ties 81
informal connections 64
informal contacts 26, 121–2
informal networks 60; benefits 148; contacts 68; emergent relations 65; Greenwood 71; informal relations 68, 69; informal ties 70; overview 132; rich ties 68–9; Siemens 45; SNL (Siemens NL) 58, 147
informal ties 64, 70, 72
information: access through connections 5; exchange 15, 16; flow 9; and social networks 5
information asymmetry 103
informed consent 140
innovation: assessing the climate 24–7; cross-hierarchy ties 80–2; cross-unit ties 80–2; despite reorganization 116; diversity 120–1; DNA 118; guardians 119–20; managing 142–5; in networks 4–8; proper network characteristics 107–9; and rich ties 73–5; social activity 4; stimulation of newcomers 109–10; sweet spot 117–18
innovation engagement scan *see* IES (innovation engagement scan)
innovation management (IM) 55–6, 59; improving the innovative climate 150–1
innovation networks 21–2, 25; overview 132
innovation roles *see* brokers
innovative project teams *see* project teams

Intel 121
internal orientation 37–9, 41; brokerage 48; brokers 42–3, 118–19, 122
interpreting data: actionable insights 138; connecting the network 'dots' 137–8; contrasting observations with past experience 138
interventions: changing behaviour 106; communication costs 106; corporate restructuring 115; crossing unit boundaries 108–9; formal 106; goal of 103; haphazard nature of 102; improved knowledge integration 105; information asymmetry 103; legitimation of certain activities 106; management 103, 104–7; network 102–4; number of contacts 107–8; proper network characteristics for innovation activities 107–9; Redrock case study 110–13; reduction of barriers to knowledge integration 105; resources and opportunity 107; sensitive nature of 102; simple 104, 104–7, 111–13, 114; stimulating newcomers 109–10; taskforce 104–7; transferring innovative knowledge 103; unintended outcomes 114
interviews: post-assessment 138–9
intra-organizational networks: nature of 3; quantity and configuration of relationships 10–11; relationships 2
isolates 131

Jive 140
Johnson, J. D. 69
joint production 142

Katila, R. 109
knowledge: accessing from across organizational boundaries 83; asymmetrical distribution of 103; available and accessible 2; combination and recombination of 5–6; complex 79; diversity of 92–3; domains 154; externally oriented 38; hubs 24; improved integration

through intervention 105; innovative 73–4; integration 105, 106; lack of awareness amongst employees and managers 26; lack of sharing 55; management of 5; specialisation at innovative team level 88; specialized 37; strategic position 7; team of experts 117
knowledge exchange 3, 15; complexity of 142; in networks 22–3; willingness 103
knowledge flow: information 9; at Siemens 44–5; at SNL (Siemens NL) 56, 58; specialization at the innovative team level 88; workflow networks 65–6
knowledge transfer: across boundaries 36; barriers 7, 36; brokerage 150; business units 37–8; cruciality of understanding of networks 5; effective 4; formal relations 74; formal structures 66; informal ties 64; innovation in organizations 65–7; innovative 73–4; multiple pathways 144; relations between employees 65; restraints 21–2; tie types 71

law of unintended consequences 17
leaders 28, 46, 121
Leenders, Roger 79, 80, 83
Liaison 40, 41, 46
linchpin 25
low density networks 108
low redundancy networks 12

managers: bringing newcomers to specific insiders 32; coalition building 16–17; creating permanent and transitory structures 17; forming project teams 123; fostering network structures 49; influencing knowledge transfer 74; intervening in formal networks 64, 74; intervention to implement innovative ideas 32, 103; nurturing innovation sweet spot 117; resources and funding 31; sense of awareness 94; shaping formal structures after downsizing 123; simple intervention 104–7; understanding of social networks 8; value of project teams 79
mandated contacts 65, 66
Mann-Whitney test 44, 51n1
Marshall, Alfred 12
matrix structure 53–4
mature markets 5
mentors 27
Merton, Robert K. 17; local and cosmopolitan network roles 39–40
Microsoft 140
Mintzberg, H.: organization charts 67
monitoring 139
Moreno, J. L. 134
multiplex combination 39
multiplex interactions *see* rich ties
multiplex networks 69–70; overview 132
multiplex ties 63, 64, 70, 72; *see also* rich ties

name generators 131–2, 133–4, 136
NBD (New Business Development) 105, 106, 107, 113
need identification: data collection 128–34; goals and objectives 127; identification of relevant stakeholders 127–8; transparency and commitment 127–8
Netflow 136
network brokers *see* brokers
network data 28
network efficiency 87–8
network intervention 102–4
network roles: categorization of 39–40; combining 40; communication profiles 40; cosmopolitan 40; local 40; *see also* communication roles
networks: diversity of 3; essence of 2; and innovative project teams 79–80; structure and position 8–15
network tactics 15–17

New Business Development (NBD) 105, 106, 107, 113
newcomers: connections 27; encouraging contributions of 32; involvement with innovation at Redrock 112; stimulating to innovate 109–10
nodes: essence of networks 2; interaction of 9; *see also* employees

objectives: need identification 127
Obstfeld, David 37
Okhuysen, G. A.: experimental study 105; simple interventions 104
opportunistic orientation 37
Ordinary Least Squares (OLS) 126
organizational charts 66–7
organizational level 28, 29
Organizational Network Analysis (ONA): analysis and visualization 136–7; applications of 23; assessing potential damage by employees 8; barriers 7; data collection 128–34; designing the questionnaire 135; features 23; IES (innovation engagement scan) *see* IES (innovation engagement scan); interpretation of data 137–40; monitoring 139; need identification 127–8; post-assessment interviews 138–9; prioritization of actions 139; research ethics 140–1; social influences 4; social structure of networks 7; survey response rates 135–6; *see also* diagnosing organizations
organizations: analysing networks 28; barriers to innovation 36; boundaries 36; boundary positions 39; constellation of networks 3–4, 7; crossing boundaries 35; dealing with complex technical knowledge 79; diagnosing *see* diagnosing organizations; formal 64; formal connections 64; formal structures 67–8, 103; informal 104; informal connections 64; informal relations 68, 69; innovative knowledge transfer 65–7; integration of new hires 27; joint production 142; micro-level processes 102–3; network tactics 15–17; organizational chart 66–7; politics of 86; pressure to innovate 4; relational embeddedness 10–11; reorganization *see* reorganization; rich ties 63–5; social complexity 4; social entities 142; social infrastructure 6; social networks 15; structural embeddedness 10, 11; understanding processes 104; value of social structures 36
orientation: brokers 118–19; external 37–9, 41, 42–3, 44, 48, 118–19; internal 37–9, 41, 42–3, 48, 118–19, 122; internal versus external 37–9, 41; opportunistic 37; strategic 37; tertius iungens 37
out-Bonachich power 30
out-closeness centrality 30
out-degree centrality 13, 30

Pajek 136
patents 53
payment methods: innovative 71
performing innovative teams 91, 92
Philips 118
politics: generating support for ideas 107; influence on project teams 82; in organizations 86
potential: unlocking 36
prioritization 139
product concepts 80
product innovation 118
project teams: access to influencers 86; accomplishing organizational objectives 78; assembling 78; bottom-up emergence 78; challenges of managing 79; collective creative insights 78; concentrated horizontal and vertical cross-ties 95–6; connected to upper

management 91; connecting horizontally 79; connecting vertically 79; external connectedness 80; external ties maintained by limited number of members 88; failure rate 80; identifying team member to establish new connections 123; interaction and cross-fertilization of ideas 83; and networks 79–80; non-routine challenges 82; political influence 82; pooling knowledge and information 78–9; pursuit of innovation 79–80; success of 84; temporary nature of 78; upward influencing 84, 85; vertical cross-hierarchy ties 85, 86; vulnerability to termination 92; *see also* cross-ties

QAP (Quadratic Assignment Procedure) analysis 76
questionnaires 135

R&D: Siemens 53
radio-frequency identification (RFID) 23
random group level 28, 29
reciprocal communication 61–2, 148–9
reciprocal relationships 9
Redrock 28; differentiated benefits of cross-ties 92–5; formal intervention 110; formal ties 72, 76; frequency of tie types 72; growth of innovation community 111–12; horizontal cross-ties 88–90; informal ties 72, 76; machine bureaucracy 70–1; meetings to emphasise relevance of cooperative behaviour 111; multiplex ties 72, 76; QAP regression 76; simple intervention 111–13; taskforce 110–11; vertical cross-ties 91–3
redundancy: in networks 12
relational embeddedness 10–11
relationship management 95–6
relationships: emergent 65; establishing and maintaining 87; formal 74; improvement between departments 25–6; informal 74; knowledge transfer between employees 65; organizationally mandated 65; quantity and configuration of 10–11; reciprocal 9
reorganization: activities to be maintained and discarded 122; contribution of well-connected employees 120; effect on employees' contributions 116; effect on social networks 116; impact on innovation 116; informal contacts 121–2; innovation DNA 118; innovation sweet spot 117–18; internally oriented brokers 118–19, 122; *see also* downsizing
Representatives 40, 41, 46
research ethics 140–1
resources: management control of 31
rich ties: formal networks 66–8; Greenwood case study 70, 71–3; informal networks 68–9; and innovation 73–5; innovative knowledge transfer in organizations 65–7; multiplexity 63, 64–5, 70; multiplex networks 69–70; in organizations 63–5; Redrock case study 70–3
Ropp, V. A.: transfer of ideas in informal networks 69
routines 106

Salesforce.com 140
sampling 130–1, 133
Schumpeter, Joseph: definition of innovation 5
Scott, W.: informal relations within organizations 68
Sharenet 55
Siemens: brokerage activity 45; business units 52; cooperation between business units 43; customer oriented approach 53; decentralized structure 52; employees with external

orientation 44; flow of information/knowledge 44–5; formal network 44; IES test 44; individuals in communication roles 46; informal network 44, 45; innovation network 46; patents 53; worldwide vision 52–3; *see also* SNL (Siemens NL)
SIENA 137
Silicon Valley 12
Simmel, G. 134
simple interventions 104, 104–7, 111–13, 114
Skvoretz, J. 65
SNL (Siemens NL): BU 1 (Transport and Distribution) 54, 55, 59, 60, 61; BU 2 (Information and Communications) 54, 55, 59, 60; BU 3 (Traffic and Safety) 54, 55, 59; BU 4 (Mobile Communications) 54; bridging boundaries 55–8; business units 54; clustering 59–61; communication flows 56; cooperation between BU 1 and BU 3 55; employees' perspectives on knowledge sharing 56–7; formal network 58, 59; incentive structures 55, 58; informal network 60; innovation management (IM) 55–6, 59, 150–1; lack of knowledge sharing 55; lack of reciprocity 148–9; matrix structure 53–4; overarching strategic goals 54; reciprocal communication 61–2; transport and mobility scheme 54
snowball procedure 131, 133
social capital 36, 42, 69
social cognition models 105
social complexity: organizations 4
social embeddedness 31
social entities 142
social infrastructure 6, 31–2; changing 102
social networks: collaboration of employees 117; data collection 128–34; impact of company reorganisation 116; and information 5; management of 6; overlapping 15; structures of 7; understanding of 3–4
social network theory 102
social structures: value to organizations 36
socio-centric sampling 130–1
Soda, G. 65, 67, 69
software programs 136
sponsorship 48, 94
stakeholders 127–8
star network 151, 152
strategic orientation 37
strategy: goals of SNL (Siemens NL) 54
strong ties 11
structural embeddedness 10, 11
structure: formal 67–8; understanding 28
subgroups 25
surveys: anonymity 135; ego-centric sampling 130, 131, 133; response rates 135–6; snowball procedure 131; socio-centric sampling 130–1
sweet spot, innovation 117–18

tactics: network 15–17
taskforces 103; formal intervention at Redrock 110; simple interventions 104–7, 111–13
TEDx 47
temporary teams *see* project teams
tertius iungens orientation 37
ties: closeness 10; cross-ties *see* cross-ties; directed 9; formal 64, 70, 72; informal 64, 70, 72; interaction of nodes 9; multiplex 63, 64, 70, 72; rich *see* rich ties; significance of 9; strength 10; strong 11; uniplex 71; weak 11, 27
transparency: need identification 127–8
transportation: cooperation between divisions 55; SNL approach to 54; solutions to problems 54
trendsetting 86
Triad based brokerage 41

under-performing innovative teams 91, 92
unintended consequences, law of 17
uniplex ties 71
unit boundaries 36; clustering of expertise 37; crossing 108–9; knowledge acquired across 38
upward influencing 84, 85

venture capitalists: and networking 5
vertical cross-ties 80, 82, 84–7, 91–3; benefits 94; concentrated 95–6

visualization 136–7

water cooler effect 90
weak ties 11, 27
well-connected employees 120
workflow networks 65–6

Yammer 140

Zaheer, A. 65, 67, 69
Zhao, M. 36